Springer Theses

Recognizing Outstanding Ph.D. Research

Aims and Scope

The series "Springer Theses" brings together a selection of the very best Ph.D. theses from around the world and across the physical sciences. Nominated and endorsed by two recognized specialists, each published volume has been selected for its scientific excellence and the high impact of its contents for the pertinent field of research. For greater accessibility to non-specialists, the published versions include an extended introduction, as well as a foreword by the student's supervisor explaining the special relevance of the work for the field. As a whole, the series will provide a valuable resource both for newcomers to the research fields described, and for other scientists seeking detailed background information on special questions. Finally, it provides an accredited documentation of the valuable contributions made by today's younger generation of scientists.

Theses are accepted into the series by invited nomination only and must fulfill all of the following criteria

- They must be written in good English.
- The topic should fall within the confines of Chemistry, Physics, Earth Sciences, Engineering and related interdisciplinary fields such as Materials, Nanoscience, Chemical Engineering, Complex Systems and Biophysics.
- The work reported in the thesis must represent a significant scientific advance.
- If the thesis includes previously published material, permission to reproduce this must be gained from the respective copyright holder.
- They must have been examined and passed during the 12 months prior to nomination.
- Each thesis should include a foreword by the supervisor outlining the significance of its content.
- The theses should have a clearly defined structure including an introduction accessible to scientists not expert in that particular field.

More information about this series at http://www.springer.com/series/8790

John Antoniadis

Multi-Wavelength Studies of Pulsars and Their Companions

Doctoral Thesis accepted by
the Rheinischen Friedrich–Wilhelms–Universität,
Bonn, Germany

Springer

Author
Dr. John Antoniadis
Max-Planck-Institut für Radioastronomie
Bonn
Germany

Supervisor
Prof. Dr. Michael Kramer
Max-Planck-Institut für Radioastronomie
Bonn
Germany

ISSN 2190-5053 ISSN 2190-5061 (electronic)
ISBN 978-3-319-38495-5 ISBN 978-3-319-09897-5 (eBook)
DOI 10.1007/978-3-319-09897-5

Springer Cham Heidelberg New York Dordrecht London

Softcover reprint of the hardcover 1st edition 2015

Printed on acid-free paper

Springer is part of Springer Science+Business Media (www.springer.com)

Each piece, or part, of the whole nature is always an approximation to the complete truth, or the complete truth so far as we know it. In fact, everything we know is only some kind of approximation, because we know that we do not know all the laws as yet. Therefore, things must be learned only to be unlearned again or, more likely, to be corrected. The test of all knowledge is experiment. Experiment is the sole judge of scientific 'truth.'

Richard Feynman
The Feynman Lectures, Introduction

For Kiki, Michalis,
Moschos, Chrysoula and
in loving memory of Andreas

Supervisor's Foreword

When a massive star explodes in a supernova, the core of the star collapses and forms an object so dense, that electrons and protons are 'squeezed into each other'. The result is a neutron star, a star with the densest matter that humankind can study. No laboratory on Earth is able to create this extraordinary material, where a tea spoon weighs as much as the largest ocean-going super-tanker. Obviously, studying these objects known as neutron stars can teach us a lot about the basic properties of matter. The obvious question is, how dense can you pack matter? Or, to express it more generally, how does the density change with pressure, or what is the relationship between mass and ratio? What is the largest amount of mass that you can pack into a neutron star? Answering these questions is difficult. Most of the information we obtain comes from radio observations of neutron stars known as pulsars.

Pulsars are rotating neutron stars that emit a radio beam along their inclined magnetic axis. This results in a lighthouse effect that causes the radiation observed on Earth to be pulsed. For a number of pulsars, the masses of neutron stars can be measured if the pulsar has a companion. In this case, the pulsar moves with the companion about the common centre of mass that leads to a systematic variation in the pulse arrival times measured on Earth. This variation in turn depends on the combination of masses of the pulsar and its companion. Usually, we need relativistic effects to determine the mass of the pulsar itself, and we have done that for quite a number of neutron stars already. However, this standard method may contain some selection effects, since it is easier to derive the masses for very compact binaries, which may have a different evolutionary history. It would therefore be good to get measurements that are independent of relativistic effects. This is the point where the thesis in your hand comes into play.

If the pulsar has a companion that can be observed in the optical, we can complement the radio information with that from optical telescopes. For instance, if the companion is a white dwarf, we can potentially measure atmospheric spectral lines that shift in frequency, as the white dwarf orbits the pulsar. The obtained radial velocity can be compared to the radial velocity of the radio pulsar, and the result is a theory-independent mass ratio of the two stars. Moreover, in many cases, we can also determine the temperature and the surface gravity of the white dwarf, which we can relate to the mass of the white dwarf. Lo and behold, both combined give us the mass of the pulsar.

While this sounds simple, the devil is in the details and it needs an extraordinary student to make this experiment work. I am very pleased to say that this thesis describes the outstanding work of such a student. John Antoniadis not only took the optical and radio data using the biggest telescopes on the planet, but he also modelled the white dwarf data, calibrated his model, and derived masses for neutron stars that were astonishing. He found the largest mass of any neutron star, and showed what this tells us about the state of super-dense matter, but also what implications that has for theories of gravity. In particular the latter aspect caused a global media frenzy. His results found their way into a wide range of media, from local TV coverage to being the most-read story in the "Wall Street Journal" (as listed on their website). While this should not necessarily be taken as marker for quality, the global attention does reflect that his research his highly relevant for a large area of physics.

I am sure that the reader will enjoy reading the thesis. The results are outstanding but also the first chapters can easily serve as a standard reference for the status of pulsar research and our knowledge. It is written extremely well.

We did not quite know what to expect when John started the thesis. At the end, it has been a remarkable journey with lots of interesting results—results so interesting that they deserve to be seen by a wide audience. It was very joyful to be a part of this journey, so that I also wish to the reader: Enjoy!

Bonn, July 2014 Prof. Dr. Michael Kramer

Abstract

With an average density higher than the nuclear saturation density, neutron stars (NSs) provide a unique test-ground for nuclear physics, quantum chromodynamics (QCD) and strong-field gravity. This thesis deals with binary pulsars with white-dwarf companions which—if modeled properly—provide an unparalleled test-ground for NS physics and gravity.

Recent improvements in observational techniques and advances in our understanding of WD interiors have enabled a series of precise mass measurements is such systems. These masses, combined with high-precision radio timing of the pulsars, result to stringent constraints on the radiative properties of gravity, qualitatively very different from what was available in the past. In addition, one of the systems presented here, J0348+0432, hosts the most massive NS known to date. This posses significant constraints on theories describing neutron-rich and ultra-dense cold matter under conditions impossible to replicate on Earth.

Acknowledgments

This document marks the end of a 3-year experience made possible by a unique group of remarkable people.

First and foremost, I would like to thank my supervisor Michael Kramer for accepting me in the "Fundamental Physics" group, for tolerating my mistakes, for always being there and, most importantly, for turning something unknown to me to an everyday experience I truly enjoy with all my heart.

Likewise, a simple thanks is not enough to express my gratitude to my advisors Paulo Freire, Norbert Wex and Thomas Tauris. I do not only consider them to be word-leading experts but also mentors and friends. Their ideas have influenced every line of this Thesis.

I would like to extent my gratitude to Marten van Kerkwijk, for teaching me everything I know about spectra, for openly sharing his expertise and enthusiasm, for supervising my work remotely, putting up with me and for answering all my questions no matter how trivial.

I thank all the members of the MPIfR pulsar group for openly sharing their expertise and for all the useful and fun discussions we had through the years. I would also like to acknowledge, the help and support of our secretary, Gabi Breuer.

I would like to thank Ryan Lynch and the rest of the J0348 fans for giving me the opportunity to work on this exciting system.

Thanks to (in semi-random order):

Maca, Miguel, Marcos, Diego, Vanessa, Giorgos, Fani, Lars, Antuanetta, Cristian, Nicolas, Kazi, Frank, Jiannis, Lilia, Jiannis, Vassilis, Natasa, Konstantina, Orestis, Eleni, Michael, Nicolas, Hananeh, Matteo, Jeff, Zahra, Marika, Aarti, Cherry, Lijing, Angela, Babafsheh, Sandra, Bia, Alice, Partick, Pablo, Silvia^2, Maria-Luisa, Senol, Pantelis, Trialda, Christos, Kostas, Maria, Maria, Stelios, Chrysa, Panagiotis, Natasa, Thanassis, Jiannis, Nikos, Kostas, Soula, Melpo, Jiannis, Zani, Dimitris, Jiannis, Olympia, Nikos, Spyros and Kostas. You have tried your best to keep me sane. I've greatly enjoyed your company, friendship and our conversations (eventhough, sometimes one-sided, given that some of you don't speak yet). Special thanks to Silvia Spezzano for unselfishly willing to lead my papers and for cooking a delicious breakfast for a hangovered homeless hippie on a cold Winter's day.

I'm obliged to Kosmas Lazaridis for his friendship and assistance during the course of this work.

Thank you Ewan Barr, among else for forcing me to search for "learn-Scottish-in-5-easy-steps" lessons on YouTube (one of the best is this one: http://www.youtube.com/watch?v=mALkCGVA2BU).

I am grateful to the International Max Planck Research School (IMPRS) for Astronomy and Astrophysics at the Universities of Bonn and Cologne for providing financial support for this research and to the IMPRS office for the outstanding quality of their work. Special thanks to Manolis Angelakis for bringing IMPRS to my attention to and for supporting me in every way possible.

I am deeply obliged to the technical staff of Effelsberg, Arecibo and VLT for their great work and professionalism.

Thanks to my family for supporting me all these years. This work is dedicated to you.

Special thanks to my family in Canada for their love, help and hospitality during my stays in Toronto. It was great to see you again after so many years.

John Seiradakis is the main person responsible for me being here.

Thank you Dhanusha Manoharan and please forgive me for not explicitly acknowledging your significant contribution to this work.

Last but not least, I thank Lia Vaikousi for her love and understanding and for always waiting at the airport.

Contents

Nomenclature

Frequently Used Symbols

b	Galactic latitude
B	Magnetic flux density
B_0	Magnetic flux density at the surface
g	Gravitational acceleration
G	Gravitational constant
c	Speed of light
δ	Declination
e	Eccentricity (or electron charge or numerical constant)
$E_{\mathrm{B-V}}$	Colour excess
Z	Metallicity (or atomic number)
h	Planck's constant
H	Hour angle
i	Inclination
J	Orbital angular momentum
k	Coulomb's constant
k_{B}	Boltzmann's constant
σ	Stefan–Boltzmann constant
λ	Longitude
L	Luminosity
μ	Mean molecular weight
m	Mass (or apparent magnitude)
M	Absolute magnitude (or mass)
ν	Frequency
n	Numerical density (or neutron)
π	Parallax (or numerical constant)

p	Momentum (or proton)
P	Pressure (or period)
q	Mass ratio
r	Distance
R	Radius
t	Time
T	Temperature
ϕ	Latitude (or angle)
v	Speed
V	Volume
X	Fraction of hydrogen
Y	Fraction of helium
Ω	Solid angle
ω	Periastron (or angular frequency)

Numerical Constants

π = 3.14156
1 rad = 57.296 degrees
e = 2.7183
$\log e = 0.4343 = \ln(10)^{-1}$

Physical Constants

Speed of light	$c = 2.9979 \times 10^{10}$ cm sec^{-1}
Gravitational constant	$G = 6.670 \times 10^{-8}$ dynes cm^2 gr^{-1}
Planck's constant	$h = 6.626 \times 10^{-27}$ erg sec
Coulomb's constant	$k = 1$
Boltzmann's constant	$k_B = 1.381 \times 10^{-16}$ erg deg^{-1}
Stefan–Boltzmann constant	$\sigma = 5.6704 \times 10^{-5}$ erg cm^{-2} s^{-1} K^{-4}
Electron mass	$m_e = 9.110 \times 10^{-28}$ gr
Proton mass	$m_p = 1.673 \times 10^{-24}$ gr

Astronomical Constants

Astronomical unit (1 AU)	$= 1.496 \times 10^{13}$ cm
Parcec (1 pc)	$= 3.086 \times 10^{18}$ cm
Julian light year (1 ly)	$= 9.460730472 \times 10^{17}$ cm
Julian year (1 yr)	$= 3.15576 \times 10^{7}$ sec
Solar mass (1 $M_\odot$)	$= 1.989 \times 10^{27}$ gr

Solar radius (1 $R_\odot$) = 6.960×10^{10} cm
Solar luminosity (1 $L_\odot$) = 3.9×10^{33} erg s^{-1}
Absolute magnitude of the Sun (M_V) = 4.77
Apparent magnitude of the Sun (m_V) = −26.7
Effective temperature of the Sun (T_{eff}) = 5770 K

Chapter 1
Neutron Stars and Pulsars

I switched on the high speed recorder and it came blip.... blip.... blip.... Clearly the same family, the same sort of stuff and that was great, that was really sweet!

Jocelyn Bell-Burnell

1.1 Birth, Life and Death

It is now understood that stellar evolution is driven by nuclear fusion, which in turn depends on the stellar mass and composition (Vogt 1926; Russell 1931), as well as on macroscopic interactions with the stellar environment. This is how it works in a nutshell:

Stars form from the gravitational collapse of fragmenting molecular clouds. During this process, a collapsing fragment of mass M, releases gravitational energy and heats-up. The thermal energy of the core nuclei opposes their mutual Coulomb repulsion and eventually forces them to trap in the attractive potential of the strong interaction:

$$\frac{3}{2}k_{\rm B}T = \frac{kZ_1Z_2e^2}{r} \tag{1.1}$$

If $r_{\rm s}$ is the range of the nuclear force, then the former condition is fulfilled when r becomes equal to the de-Broglie wavelength, $r = \lambda = h/p = (h^2/3mk_{\rm B}T)^{1/2}$, increasing the chance for a close encounter through quantum tunnelling (which scales as $e^{-r_{\rm s}/\lambda}$). From Eq. 1.1 it immediately follows that the critical temperature for this is:

$$T = \frac{4}{3}\frac{k^2Z_1^2Z_2^2e^4m}{k_{\rm B}^2h^2} \tag{1.2}$$

J. Antoniadis, *Multi-Wavelength Studies of Pulsars and Their Companions*, Springer Theses, DOI 10.1007/978-3-319-09897-5_1

Equation 1.2 implies that the *first* element to ignite is hydrogen (when $T \sim 10^7$ K), which also happens to be the most abundant. The released energy of the exothermal reaction balances the gravitational attraction and brings matter to thermal equilibrium; the star is on the main sequence. Main-sequence stars with masses smaller than $\sim$1.5 $M_\odot$ primarily fuse their hydrogen through the proton-proton (pp-) chain while more massive ones through the CNO bi-cycle.[1]

When the core's hydrogen is exhausted, equilibrium breaks, and the star contracts again. For masses below $M \sim 0.25\,M_\odot$, the temperature never becomes high enough to re-ignite the core (now made of helium), and equilibrium is reached only due to electron Fermi gas pressure (discussed below), transforming the star into a *helium-core white dwarf*. For $M > 0.25\,M_\odot$, the burning process continues with helium fusion after the star has climbed the Red-Giant branch on the H-R diagram. Long-story short, the cycle of contraction and re-ignition continues and sufficiently massive stars form white dwarfs with cores made of progressively heavier elements (carbon, oxygen etc.).

A critical situation arises when the stellar core grows beyond $\sim$1.2–1.5 $M_\odot$ and, for whatever reason, the burning reactions halt. To understand what happens, it is necessary to recap the physics of a self-gravitating Fermi gas.

1.1.1 Fermi Gasses and the Chandrasekhar Limit

Let us consider a (sufficiently small) sphere of radius R that contains a (sufficiently large) number of fermions, say protons, neutrons and electrons, that obey the Fermi-Dirac statistics ($T \to 0$). The central inward pressure exerted due to gravity is

$$P_{\rm g} = -\frac{3}{8\pi}\frac{GM^2}{R^4}, \tag{1.3}$$

where $M = \int_0^R 4\pi\rho_0 r^2\,dr$ is the total mass and the density ρ_0 is taken to be uniform over the sphere. We need now find the opposing-outward pressure caused by the degenerate gas. The number-density per-unit-volume of identical fermions with momenta between p and $p + dp$ is

$$n(p)dp = F(p)\frac{g4\pi}{h^3}p^2dp, \tag{1.4}$$

where $F(p)$ is the Fermi-Dirac distribution function, which reads $F(p) = 1$ for $p < p_{\rm F}$ and $F(p) = 0$ for $p > p_{\rm F}$, and $g = 2$ for fermions. We can now find the Fermi momentum for a given particle density, n by integrating Eq. 1.4 to infinity:

[1] Obviously these masses depend on the availability of CNO catalysts and therefore on the metallicity.

$$n = \int_0^{\infty} n(p)dp = \frac{8\pi}{h^3} \int_0^{p_{\mathrm{F}}} p^2 dp \Rightarrow p_{\mathrm{F}} = \left(\frac{3h^3}{8\pi} n \right)^{1/3}. \tag{1.5}$$

The pressure P exerted by the gas, assuming that it behaves ideally is $P = \frac{1}{3} n \langle pu \rangle$ which, combined with the above, in the relativistic limit yields:

$$P = \frac{8\pi}{3h^3 m} \int_0^{p_{\mathrm{F}}} \frac{p^4}{\sqrt{1 + p^2/m^2c^2}} dp. \tag{1.6}$$

With a change of variables, $x = p_{\mathrm{F}}/mc$, the integration gives:

$$P = \frac{\pi m^4 c^5}{3h^3} \left\{ x\sqrt{1+x^2}(2x^2 - 3) + 3 \ln \left[x + \sqrt{1+x^2} \right] \right\} \tag{1.7}$$

In the classical limit ($p_{\mathrm{F}} \ll mc, x \rightarrow 0$), Eq. 1.5 reduces to:

$$P = \frac{1}{20} \left(\frac{3}{\pi} \right)^{2/3} \frac{h^2}{m^2} n^{5/3}, \tag{1.8}$$

while in the fully relativistic limit ($x \rightarrow \infty$) one gets:

$$P = \frac{1}{8} \left(\frac{3}{\pi} \right)^{1/3} hcn^{4/3}. \tag{1.9}$$

Now that we have a feeling for the equation-of-state (EoS, $P(n)$) we can calculate the maximum mass that can be supported by this pressure. Assuming that our ideal ball contains approximately equal number of neutrons, protons and electrons (so that it is overall neutral) and that the exerted pressure is only due to the electron gas, the condition $P < P_{\mathrm{g}}$ yields:

$$M_{\mathrm{max}} \leq \frac{3}{16\pi} \left(\frac{hc}{G} \right)^{3/2} \frac{1}{\mu_{\mathrm{e}^-} m_n^2}, \tag{1.10}$$

where $\mu_{e^-} = A/Z \sim 2$ is the mean molecular weight per electron. M_{max} gives the maximum mass of a body that can be supported by electron pressure. It was discovered by Chandrasekhar during his first boat trip to the U.K., on his way to become a graduate student.

Modern calculations that take into account low-pressure and finite-temperature corrections place the "Chandrasekhar Mass" limit around $\sim$1.3 $M_{\odot}$. Any star capable of developing a core beyond M_{max} is a candidate for giving birth to a neutron star or a black hole. During the transition, things may become explosive.

1.2 Supernovae and the Birth of Neutron Stars

1.2.1 Core-Collapse Supernovae

In the textbook example of core collapse, stars with masses above $\sim$6–11 $M_\odot$ ultimately develop massive Silicon $^{28}_{14}$Si cores. The latter fuse to $^{60}_{30}$Zn which is unstable and rapidly decays to $^{56}_{28}$Ni and finally $^{52}_{26}$Fe. These nuclei of the iron group are stable (Wapstra and Audi 1985) and do not transform into heavier elements. When the mass grows beyond the Chandrasekhar limit and fusion ceases, electron pressure fails to counteract gravity and the star collapses.

Stars with initial masses above $\sim$40 $M_\odot$ probably form directly a black hole (Heger et al. 2003). For lower-mass stars, the extreme pressure raises the central temperature to some 10^{11} K and produces a thermal spectrum that peaks at γ-rays. The released energy results in photo-disintegration of the iron nuclei to neutrons and protons. It also favours electron capture from the protons via inverse β-decay ($p + e^- \rightarrow n + \nu_e$), which results in *neutronization* of matter. The emitted neutrinos interact with the stellar envelope—which in the meantime has bounced on the core and is moving outwards, and accelerate it dramatically (Woosley and Janka 2005; Janka et al. 2007). The released energy results in an electromagnetic counterpart that outshines the hole galaxy.

The former process, called a *core-collapse supernova*, marks the birth of a neutron star: a ball of neutrons $\sim$20 km wide that has a mean density higher than that of an atomic nucleus. The most massive neutron stars of neutron stars formed via this channel collapse further into black holes, after fall-back of material onto their surface (Fig. 1.1). The mass threshold for that process depends strongly on the mater equation-of-state in neutron star interiors, the detailed description of which remains elusive and is further discussed below.

1.2.2 Electron-Capture Supernovae et al.

While the overwhelming majority of neutron stars form in core-collapse supernovae, some of them also emerge from lighter stars, owing to loss of outwards or increase of inwards pressure:

- Stars with initial masses between $\sim$9 and 10 $M_\odot$ ultimately develop Oxygen-Neon-Magnesium cores, more massive than the Chandrasekhar limit. When the remaining fuel is exhausted, pressure neutralizes the nuclei (Miyaji et al. 1980; Iben and Renzini 1983). The series of events is very similar to what described above and results in an *electron-capture supernova* (Heger et al. 2003). The neutron stars formed via this channel are tentatively thought to be less massive and have different spin properties than their direct core-collapse counterparts. Furthermore, there is increasing evidence that neutron stars formed via this channel receive small super-nova kicks (Knigge et al. 2011).

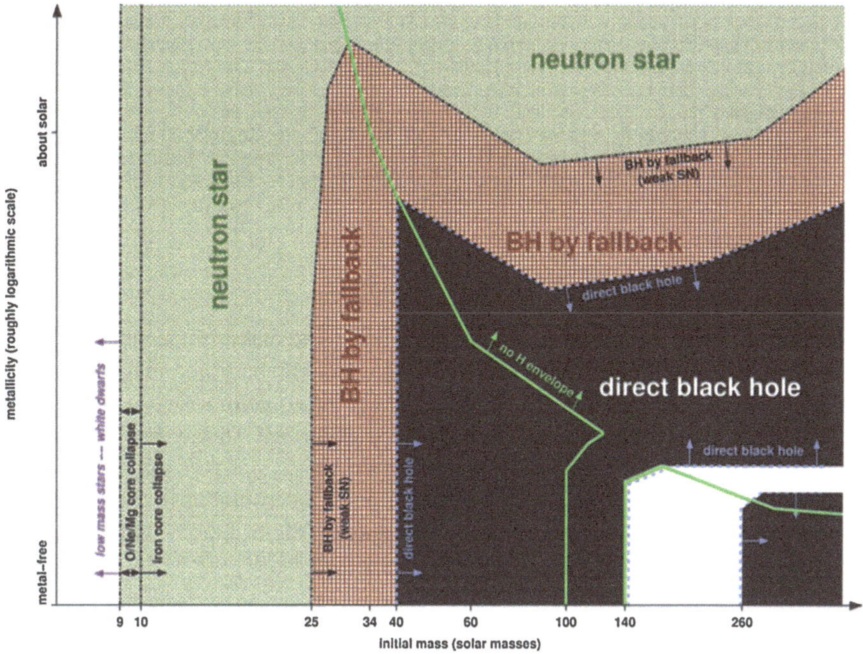

Fig. 1.1 Stellar remnants as a function of initial stellar mass and metallicity. Taken from Heger et al. (2003)

- Electron capture can be induced by transfer of mass onto a lighter-than-9 $M_\odot$ progenitor, that would normally descent to a O/Ne/Mg-core white dwarf (Dessart et al. 2006). This process is called *accretion-induced collapse* and is thought to be responsible for a large fraction of neutron stars in dense stellar environments.

Regardless of their formation channel, neutron stars count among the most extreme objects that can exist in the Universe. In the remaining of this chapter I briefly discuss the salient properties that give them their fame.

1.3 Neutron Star Structure

Neutron stars were devised on paper long before their actual discovery. Soon after the firm detection of neutrons in Sir James Chadwick's laboratory, Walter Baade and Fritz Zwicky proposed that super-nova explosions "represent the transition of an ordinary star to a neutron star" (Baade and Zwicky 1934a, b, c). Some years later, Tolman, Oppenheimer and Volkov (TOV), derived the equations of hydrostatic equilibrium in General Relativity (Tolman 1939; Oppenheimer and Volkoff 1939), necessary for inferring the structure and behaviour of these stars. These read:

$$\frac{dm}{dr} = 4\pi r^2 \rho, \tag{1.11}$$

$$\frac{dP}{dr} = -\frac{G\rho m}{r^2}\left(1+\frac{P}{\rho c^2}\right)\left(1+\frac{4\pi P r^3}{mc^2}\right)\left(1-\frac{2Gm}{rc^2}\right)^{-1}, \tag{1.12}$$

$$\frac{dm_{\mathrm{B}}}{dr} = \frac{4\pi \rho r^2}{\sqrt{1-2Gm/rc^2}}. \tag{1.13}$$

Integration of Eq. 1.11 gives the inertial mass (baryonic mass minus the negative of the gravitational binding energy) and Eq. 1.13 yields the baryonic mass. The TOV system is open and needs explicit information for the equation-of-state, $P = P(\rho)$ to be solved. The latter depends on the nature of strong interactions at densities up to ten times the nuclear-saturation density ($\rho_0 \sim 2.7 \times 10^{14}\,\mathrm{g\,cm^{-3}}$; Shapiro and Teukolsky 1969). This is far beyond the energies available in the controlled environment of terrestrial experiments. Consequently the equation-of-state remains highly uncertain. Below we discuss what is known so far.

1.4 Equation-of-State

Neutron Star Crust For areas close to the atmosphere, one can assume that the pressure approaches the ambient value and thus the equilibrium nucleus is $^{56}\mathrm{Fe}$, which has the highest binding energy per nucleon. At very small depths the iron nuclei form a lattice which is surrounded by an electron cloud, like in earthly conditions (Carroll and Ostlie 1996). Further below, pressure is provided by the degenerate electron gas as we discussed above: $P = K_1 n^{5/3}$ and then $P = K_2 n^{4/3}$.

Neutronization As we progressively move towards higher depths and densities, the conditions favour heavier neutron-rich isotopes: When the pressure raises at about half the nuclear-saturation density, $\rho \sim 0.5\rho_0$, the Fermi Energy becomes higher than the mass difference between neutrons and protons, $E_{\mathrm{F}} > \Delta E = (m_{\mathrm{n}} - m_{\mathrm{p}})c^2 = 1.29\,\mathrm{MeV}$, and the equilibrium of β-decay is shifted towards higher neutron concentrations. According to the Saha equation,

$$\frac{n_{\mathrm{p}} n_{\mathrm{e}}}{n_{\mathrm{n}}} = \frac{Z_{\mathrm{p}} Z_{\mathrm{e}}}{Z_{\mathrm{n}}}. \tag{1.14}$$

where Z_{i} is the partition function of each component, which can be broken into the product of an internal energy factor and a kinetic factor (Carroll and Ostlie 1996). After a change of variables, $n_{\mathrm{p}} = n_{\mathrm{e}} = xn$, $n_{\mathrm{n}} = (1-x)n$, the former equation yields:

$$\frac{x^2}{1-x} = \frac{8\pi m k_B T}{n h^3}\left(2 p_F e^{\frac{-p_F}{2mk_BT}} + \sqrt{2\pi m k_B T}\,\mathrm{erfc}\left(\frac{p_F}{\sqrt{2mk_BT}}\right)\right) e^{\Delta E}. \quad (1.15)$$

Here, $m = m_p m_e / m_n$ and we have assumed that the particle momenta can take any value from p_F to infinity. As p_F increases, $x \rightarrow 0$ and neutrons dominate; hence the name "neutron star".

Neutron Drip and Superfluidity Although neutrons are initially formed within the nucleus, with increasing pressure strong interactions favour a phase transition to an unbound state. This happens above $\rho_{drip} \sim 4 \times 10^{11}\,\mathrm{g\,cm^{-3}}$ which is called the neutron drip line (Shapiro and Teukolsky 1969). At the nuclear-saturation density almost all matter is transformed to free floating neutrons. This fluid has the properties of a superfluid with no viscosity. Rotation breaks the latter into vortices that make the crust rotate rigidly.

The Core Above the nuclear-saturation density the composition is largely unknown and the behaviour of matter depends strongly on the nature of the short-range interactions between the particles (Shapiro and Teukolsky 1969; Lattimer and Prakash 2007). If these allow situations where the available energy density is $\gtrsim$140 MeV, the neutrons will start emitting pions via $n \rightarrow \pi^- + p^+$. These will then form a Bose-Einstein condensate with many particles at the minimum energy that do not contribute to the overall pressure. An attractive alternative possibility is the excitation of quark degrees of freedom, i.e. strange matter composed of free quarks not confined in nuclei. These so-called *soft* equations-of-state allow for larger pressures and generally result to stars with smaller masses compared to baryon equations-of-state (Lattimer and Prakash 2007).

The determination of the underlying (correct) behaviour of matter at such high densities is an open problem that can only be probed with neutron-star observations.

1.4.1 Tackling the Equation-of-State Problem

But how can we probe the dense-matter physics using neutron stars? For a given equation-of-state and a given central density, integration of Eqs. 1.11 and 1.12 yields a star of a specific mass and radius. For different initial conditions (i.e. central densities) the solutions form a continuous line on the mass-radius plane with a one-to-one correspondence to the underlying equation-of-state (Fig. 1.2). This correlation opens a window for experimental constraints.[2]

Ideally, one would aim to measure both mass and radius simultaneously and for a range of different neutron stars. Unfortunately, this has so-far been achieved only for a handful of weakly-magnetized neutron stars undergoing nuclear-powered X-ray bursts as a result of accretion from their Roche-lobe filling companion

[2] Strictly speaking, neutron star interiors can be probed without a-priori knowledge of the mass and radius, e.g. through pulsar-glitch and seismology. However, these methods will not be considered here [see Lattimer and Prakash (2007) for an excellent review].

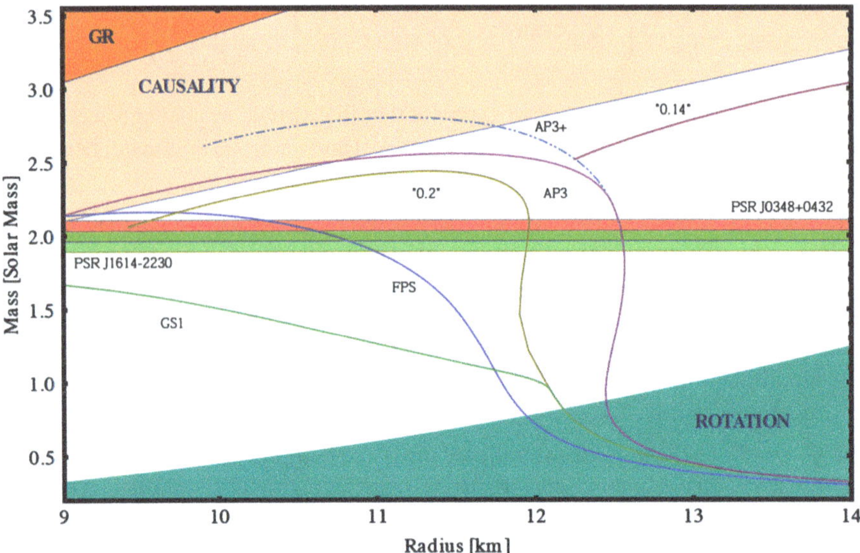

Fig. 1.2 Example mass–radius relations of neutron stars for equations-of-state with different degrees of "softening" due to matter phase-transitions. Soft equations-of-state, such as GS3, generally predict smaller maximum masses and are therefore excluded by observations. On the other hand, stiffer equations-of-state (FPS, AP3, AP3+, "0.2" and "0.14") yield larger maximum masses, consistent with the experimental constraints. Note that radii larger than ~13 km seem to be inconsistent with the constraints from observations of type-1 thermonuclear X-ray bursts (Steiner et al. 2010) and are therefore also excluded (these constraints are not shown here). The TOV system has been solved using a Runge-Kutta method of the 4th order. Tabulated equations-of-state were taken from Haensel et al. (1981) ("0.2" and "0.14") and Lattimer and Prakash (2001) (FPS, AP3, GS3). In all cases, the "SLy" equation-of-state (Douchin and Haensel 2001) was used for the outer core and inner crust and further connected to the laboratory-based values of Haensel and Pichon (1994) for the outer crust. The AP3+ curve was produced by fitting the AP3 equation-of-state using a piecewise polynomial with 3 fixed fiducial densities ($\rho_3 = 2\rho_2 = 2\rho_1 = 1.85 \times 10^{14.3}\,\mathrm{g\,cm^{-3}}$, Read et al. 2009a; Özel and Psaltis 2009) and subsequently increasing by a factor-of-two the pressure at ρ_3, thereby allowing for a higher maximum mass. Over-plotted are the constraints imposed by general relativity, causality, the two most massive neutron stars known (see text) and the fastest-spinning pulsar (716 Hz, Hessels et al. 2006)

(e.g. Özel et al. 2010). Despite that a large number of such observations are now available (thanks to sensitive X-ray satellites such as the Rossi X-ray Timing Explorer mission), masses and radii have been inferred only in a handful of occasions, due to the general lack of accurate distance information (Heinke et al. 2006; Özel et al. 2010). Furthermore, even for these exceptional cases, the large systematic uncertainties (Steiner et al. 2010) do not allow for definite conclusions.

An alternative (and less complicated) method relies on the mass measurement alone (Lattimer and Prakash 2007). This is possible in a variety of occasions, e.g. when the motion of the neutron star can be studied in a binary (see next chapter). First, when the mass of a neutron star is known, we can constraint its radius by requiring that the star is larger than its Schwarzschild radius $R \geq 2GM/c^2$. An even more stringent constraint is set by requiring causality, i.e. that the speed-of-sound is always smaller than the speed-of-light: $v_s = (dP/d\rho)_S^{1/2} \leq c$, where S is the entropy per-baryon. Over the hole volume this translates to: $R \geq 2.83GM/c^2$ (Lattimer and Prakash 2007).

In addition to the theory-independent constraints, each proposed equation-of-state has to be able to support a star *at least* as massive as the most massive neutron star observed (Fig. 1.2). Figure 1.3 summarizes all available mass measurements by the time of writing. With a few glaring exceptions, most observed neutron stars have masses of 1.2–1.4 $M_\odot$. This range is not very constraining since most equations-of-state, including those of "strange quark stars", are consistent with this limit. The situation recently changed with the accurate measurement of a 2 $M_\odot$ neutron star (Demorest et al. 2010) in the binary PSR J1614−2230. Its mass is high enough to exclude "free floating" quarks in the core (Özel et al. 2010) and suggests that if quarks do exist, they have to be strongly-interacting and colour-conducting. Currently, the most massive neutron star with a precise mass measurement known is PSR J0348+0432, presented in Chap. 5. Evidence for even more massive neutron stars has recently been found by van Kerkwijk et al. (2010) and Romani et al. (2012) for two pulsars with low-mass companions. However, due to systematic uncertainties arising from the complicated phenomenology of these systems, the validity of these measurements remains to be seen (Fig. 1.3).

Finally, information about the rotational period sets an additional constraint because the surface velocity has to be lower than the break-up velocity. For a given equation-of-state the maximum spin is allowed for the star with the highest central density, which means that, roughly, one can relate this quantity to the maximum possible mass. In fully relativistic calculations (e.g. Stergioulas and Friedman 1995), this limit is given by $(\Omega_{\rm max}/10^4\,{\rm s}^{-1}) \simeq \kappa (M_{\rm NS}/M_\odot)^{1/2}(R_{\rm NS}/10\,{\rm km})^{-3/2}$. Here κ is a numerical constant which ranges from $\kappa = 0.77$ (Haensel and Zdunik 1989) to $\kappa = 0.786 \pm 0.030$ (Read et al. 2009b) for most equations-of-state. All constraints imposed by current observations are shown in Fig. 1.2.

Looking at Fig. 1.3 we notice a complete correlation between accurate mass measurements and radio pulsars, a unique manifestation of neutron stars with clock-like properties comparable with our best atomic clocks. For the remaining of this chapter (and thesis) we will focus on these remarkable objects and the ways they can be used to probe fundamental physics.

1.5 Pulsars

Neutron stars were recognised as part of reality in the late 1960s, after Jocelyn Bell and Antony Hewish picked up an unusual radio signal with their high temporal resolution telescope. Unlike anything else detected before, the signal was highly periodic ($P = 1.33$ s), dispersed in frequency and kept sidereal time. The subsequent discovery of three similar sources and confirmation with other telescopes excluded the

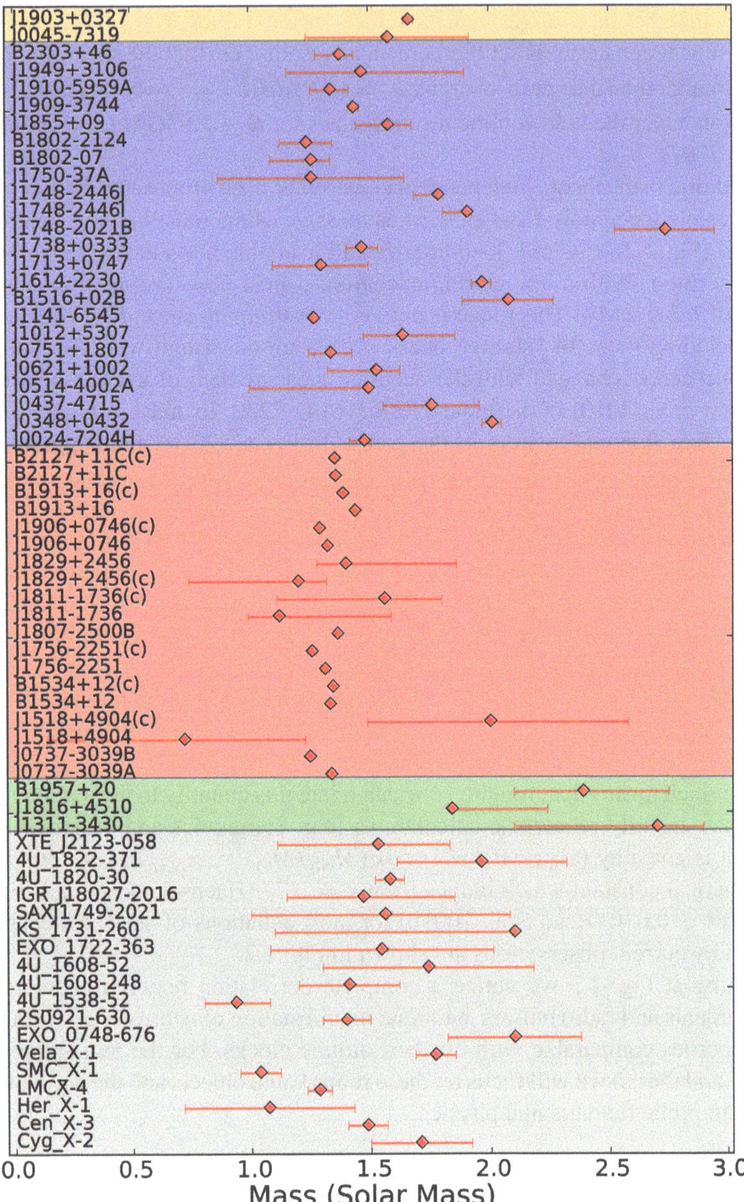

Fig. 1.3 Measured masses of neutron stars by the time of writing (style adopted from Lattimer and Prakash 2007). Colours depict different types of NS systems. From *top* to *bottom* Main-sequence/neutron-star binaries (*yellow*), white-dwarf/neutron star binaries (*blue*), double neutron stars (*red*), "*black widows*" and "*redbacks*" (*green*) and X-ray binaries. Links to the original papers can be found at http://jantoniadis.wordpress.com/research/ns-masses/

possibility for terrestrial origin. Follow-up observations also ruled out most proposed (reasonable) explanations: The signal was fast—thus the source of origin had to be small, and displayed no apparent irregularities—thus it also had to be large. Pacini (1967) and Gold (1968) proposed that the regularity of the signals can be explain if one accepts that they originate from a spinning neutron star powered by a magnetic field: with its ~10 km radius, fast rotation would not pose a stability problem and with its high moment of inertia, spinning down would require a huge amount of energy. The pulsed signals would then be instances of beamed emission from the surrounding plasma as it sweeps the line of sight of our telescopes. Indeed, the discovery of the 33 ms Crab Pulsar and the measurement of its spin-period derivative confirmed the Pacini and Gold predictions and solidified their model.

1.5.1 Pulsar Emission

The details of pulsar emission are, until today, poorly understood. However, we have good reasons to believe that their basic properties would be the same if their external magnetic field was a pure magnetic dipole of the form

$$B = \frac{\mu_0}{4\pi}\left(\frac{3\mathbf{r}\,(\mathbf{m}\,\mathbf{r})}{r^5} - \frac{\mathbf{m}}{r^3}\right), \tag{1.16}$$

where $|\mathbf{m}| = B_0/R^3$ is the magnetic dipole moment and $|\mathbf{r}| \geq R$ (Fig. 1.4). In the most general case the magnetic field and spin axes are misaligned by an angle θ.

We shall first consider a pulsar where the spin and magnetic axes are parallel. Assuming conservation of magnetic flux during the super-nova collapse, B_0 has to be at least $\sim 10^9$ G. The field rotation induces a strong electric field that strips off charge particles from the surface. Consequently, the surrounding plasma builds-up in density until it reaches an equilibrium state in which the plasma-induced electric field cancels out that from the neutron star. This force-free state allows the charges to co-rotate rigidly with the star, up to a radius $r_c = Pc/2\pi$ where the speed of the particles equals that of light. At that point (a.k.a. *the light cylinder radius*), the

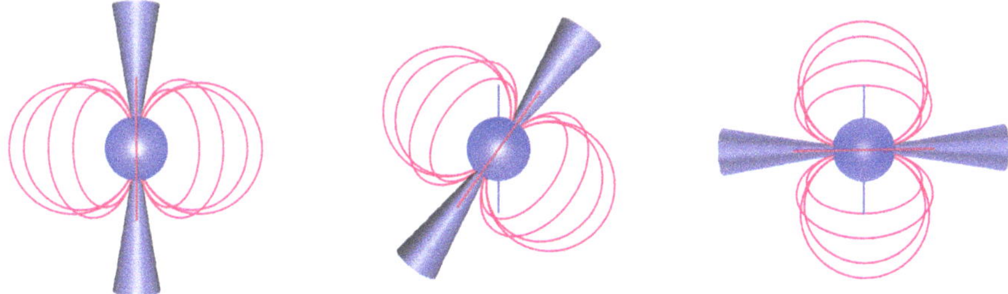

Fig. 1.4 Dipole magnetic field of a pulsar for $\alpha = 0, 45$ and $90°$

magnetic field lines are forced open and any particle trapped along them accelerates due to the large potential around r_c (Lorimer and Kramer 2005). The geometrical area of open-field lines forms a cone around the magnetic pole known as the *polar cap* region which is thought to be the source of beamed emission.

1.5.2 Spin-Down and Ages

Pulsars have been observed to spin down, typically at rates $\dot{P} \sim 3\,\mu$s per century. This results to a loss of rotational kinetic energy at a rate of (Lorimer and Kramer 2005):

$$\dot{E}_{\rm rot} = -I\Omega\dot{\Omega} = 4\pi^2 I \dot{P} P^2 \tag{1.17}$$

where I is the moment of inertia and Ω the angular velocity.

If we *assume* that the kinetic energy is carried away in the form of electromagnetic radiation induced by the spinning magnetic dipole, then

$$\dot{E}_{\rm spin} = -I\Omega\dot{\Omega} = \dot{E}_{\rm dipole} = \frac{2}{3c^3}|\mathbf{m}|^2\Omega^4 \sin^2\alpha \tag{1.18}$$

where we have considered the most general case were the magnetic and spin axes are misaligned. A rearrangement of Eq. 1.18 yields:

$$B = 3.2 \times 10^{19}\sqrt{\frac{P\dot{P}}{\sin^2\alpha}}\ \mathrm{G}. \tag{1.19}$$

This equation allows to estimate the age of the pulsar, assuming a spin-down law of the form $\dot{\nu} = \dot{P}^{-1} = -K\nu^n$:

$$T = -\frac{\nu}{\dot{\nu}}\left[1 - \left(\frac{n}{n_0}\right)^{n-1}\right], \tag{1.20}$$

where ν_0 is the initial spin frequency of the pulsar and $n = 3$ for a pure magnetic dipole. If we further assume that the pulsar was spinning much faster when it was born the above reduces to the simple

$$T = -\frac{\nu}{2\dot{\nu}} = \frac{P}{2\dot{P}} \tag{1.21}$$

called the Characteristic Age (Lorimer and Kramer 2005).

1.5.3 The P–$\dot{P}$ Diagram and Binary Pulsars

Figure 1.5 shows the distribution of spin periods and their derivatives for all known pulsars in the Galactic disk. While most of them cluster around $P \simeq 1\,\mathrm{s}$ and $\dot{P} \simeq 10^{-15}\,\mathrm{s\,s^{-1}}$, some 100 of them have millisecond periods and very small spin period derivatives. The overwhelming majority of millisecond pulsars are found in orbit around low- or intermediate-mass white dwarfs. Their fast spin periods are thought to be the result of mass accretion from the progenitor of the white dwarf (Alpar et al. 1982). This "recycling" process transfers mass and angular momentum to the pulsar and, through a mechanism not yet understood, buries its high magnetic field. Millisecond pulsars show extraordinary rotational stability and therefore make excellent probes of their environments. In binary systems, their clock-like periodic pulses can be used to infer their orbital motion with high precision. Because of their exotic nature, the orbital characteristics can then be used to map intrinsic neutron-star properties and their influence on their surroundings. This will be the subject of the following chapters.

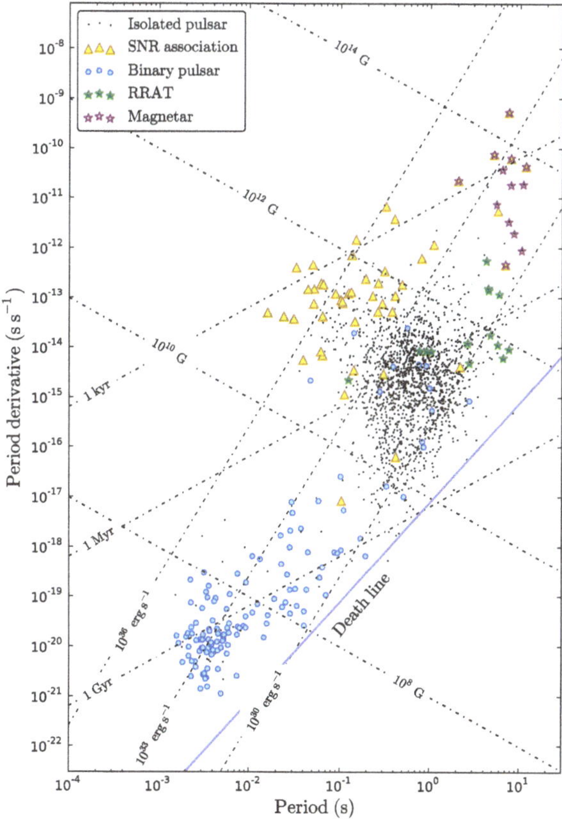

Fig. 1.5 A $P\dot{P}$–diagram of the 1805 known radio pulsars in the Galactic disk (adapted version of the same figure in Ewan Bahr's Ph.D. thesis). *Colours* depict different pulsar population. *Dotted lines* correspond to example values for quantities described in the text. Finally, the *blue* shows the "death line", i.e. the critical period at which pulsar emission ceases (Chen and Ruderman 1993)

1.6 Thesis Outline

This thesis deals with optical, radio and theoretical studies of a selected sample of binary pulsars with white dwarf companions. The text is organized as follows:

- Chapter 2 begins with discussing the physics of pulsar–white dwarf binaries: the way they come to life, their properties and their use as laboratories for fundamental physics, such as the equation-of-state of dense matter and strong-field gravity. The last part of the chapter focuses on strong-field gravity. In particular, I discuss how the motion of a pulsar in a binary is sensitive to deviations from General Relativity, even if these vanish in the Solar System and in other astrophysical objects.
- Chapter 3 describes spectroscopic and optical observations of the low-mass white-dwarf companion to PSR J1909−3744. For this system, radio-observations have yielded a precise mass measurement as well as distance information. Combined with the optical data, these provide the first observational test for theoretical white-dwarf cooling models and spectra. The latter, if correct, can be used to infer the masses of similar systems, independently of strong-field effects.
- In Chap. 4, I discuss the measurement of the component masses in the short-orbit PSR J1738+0333 system based on spectroscopy of its white-dwarf companion. This system is particularly important for understanding the physics of pulsar recycling and binary evolution. Moreover, combined with the measurement of the orbital decay from radio-timing, the masses pose the most stringent constraints for a wide range of scalar-tensor gravity theories.
- Chapter 5 describes radio and optical observations of PSR J0348+0432, an ultra-compact pulsar–white dwarf binary discovered recently with the 100-m Green-Bank Radio Telescope. Spectral observations of its bright white-dwarf companion show that the neutron star in the system is the most massive known to date. This measurement is based on a new set of white-dwarf cooling models, designed to take into account the remaining uncertainties not constrained by PSR J1909−3744. Furthermore, I discuss radio-timing observations that yield a significant measurement of the orbital decay which is completely consistent with the General-Relativity prediction. This provides a verification of the theory in a highly non-linear gravitational regime, far beyond the reach of previous experiments. PSR J0348+0432 also poses the most stringent constraints on the equation-of-state at supra-nuclear densities and sheds light to the evolution of low-mass X-ray binaries.
- Chapter 6 describes the detection of the optical counterpart of the 1 $M_\odot$ companion to PSR J1141−6545 that verifies its white dwarf nature. This simple observation is particularly important for understanding the unique evolutionary history of the binary and also verifies the constraints on alternative gravity theories imposed by the system, which up to now were based on less convincing arguments.
- Finally, Chap. 7 summarizes the main conclusions of this work.

References

Alpar M. A., Cheng A. F., Ruderman M. A., Shaham J., 1982, Nature, 300, 728
Baade W., Zwicky F., 1934a, Physical Review, 45, 138
Baade W., Zwicky F., 1934b, Proceedings of the National Academy of Sciences, 20, 259
Baade W., Zwicky F., 1934c, Proceedings of the National Academy of Sciences, 20, 254
Carroll B. W., Ostlie D. A., 1996, An Introduction to Modern Astrophysics
Chen K., Ruderman M., 1993, ApJ, 408, 179
Demorest P. B., Pennucci T., Ransom S. M., Roberts M. S. E., Hessels J. W. T., 2010, Nature, 467, 1081
Dessart L., Burrows A., Ott C. D., Livne E., Yoon S.-C., Langer N., 2006, ApJ, 644, 1063
Douchin F., Haensel P., 2001, A&A, 380, 151
Gold T., 1968, Nature, 218, 731
Haensel P., Pichon B., 1994, A&A, 283, 313
Haensel P., Proszynski M., Kutschera M., 1981, A&A, 102, 299
Haensel P., Zdunik J. L., 1989, Nature, 340, 617
Heger A., Fryer C. L., Woosley S. E., Langer N., Hartmann D. H., 2003, ApJ, 591, 288
Heinke C. O., Rybicki G. B., Narayan R., Grindlay J. E., 2006, ApJ, 644, 1090
Hessels J. W. T., Ransom S. M., Stairs I. H., Freire P. C. C., Kaspi V. M., Camilo F., 2006, Science, 311, 1901
Iben Jr. I., Renzini A., 1983, ARA&A, 21, 271
Janka H.-T., Langanke K., Marek A., Martínez-Pinedo G., Müller B., 2007, Phys. Rep., 442, 38
Knigge C., Coe M. J., Podsiadlowski P., 2011, Nature, 479, 372
Lattimer J. M., Prakash M., 2001, ApJ, 550, 426
Lattimer J. M., Prakash M., 2007, Phys. Rep., 442, 109
Lorimer D. R., Kramer M., 2005, Handbook of Pulsar Astronomy. Cambridge University Press
Miyaji S., Nomoto K., Yokoi K., Sugimoto D., 1980, PASJ, 32, 303
Oppenheimer J. R., Volkoff G. M., 1939, Physical Review, 55, 374
Özel F., Baym G., Güver T., 2010, Phys. Rev. D, 82, 101301
Özel F., Psaltis D., 2009, Phys. Rev. D, 80, 103003
Özel F., Psaltis D., Ransom S., Demorest P., Alford M., 2010, ApJ, 724, L199
Pacini F., 1967, Nature, 216, 567
Read J. S., Lackey B. D., Owen B. J., Friedman J. L., 2009a, Phys. Rev. D, 79, 124032
Read J. S., Lackey B. D., Owen B. J., Friedman J. L., 2009b, Phys. Rev. D, 79, 124032
Romani R. W., Filippenko A. V., Silverman J. M., Cenko S. B., Greiner J., Rau A., Elliott J., Pletsch H. J., 2012, ApJ, 760, L36
Russell H. N., 1931, MNRAS, 91, 951
Shapiro S. L., Teukolsky S. A., 1969, Black holes, white dwarfs, and neutron stars: The physics of compact objects
Steiner A. W., Lattimer J. M., Brown E. F., 2010, ApJ, 722, 33
Stergioulas N., Friedman J. L., 1995, ApJ, 444, 306
Tolman R. C., 1939, Physical Review, 55, 364
van Kerkwijk M. H., Rappaport S. A., Breton R. P., Justham S., Podsiadlowski P., Han Z., 2010, ApJ, 715, 51
Vogt H., 1926, Astronomische Nachrichten, 226, 301
Wapstra A. H., Audi G., 1985, Nuclear Physics A, 432, 1
Woosley S., Janka T., 2005, Nature Physics, 1, 147

Chapter 2
Binary and Millisecond Pulsars

No, I don't understand my husband's theory of relativity, but I know my husband, and I know he can be trusted.

Elsa Einstein

2.1 The Observed Population of Binary Pulsars

The first evidence for neutron stars residing in binaries came in the early 60s when Giacconi et al. (1962) discovered the first extrasolar X-ray source, Sco X$-$1. Its high X-ray luminosity of $L > 10^{35}\,\mathrm{erg\,s^{-1}}$ could be naturally understood if the source is powered by a compact object (neutron star or black hole) that accretes mass from a stellar companion. Matter falling onto the surface would then result in significant release of gravitational energy which, due to the small column density of $\sim 0.3\,\mathrm{g\,cm^{-2}}$, can easily penetrate the system in the form of X-rays (Tauris and van den Heuvel 2003). This hypothesis was confirmed with the discovery of 4.9 s pulsations from the 2 d binary Cen X$-$3 (Schreier et al. 1972).

The first binary radio pulsar was discovered some years after by Hulse and Taylor (1975) during a sensitive survey conducted with the 300-m Arecibo radio telescope (Hulse and Taylor 1974). The "Hulse–Taylor" binary consists of two neutron stars (one of them is the pulsar) that orbit each other every 7.75 h. Among else, radio-timing observations yielded the first accurate determination of neutron-star masses and the first indirect detection of gravitational waves through the measurement of the system's orbital decay.

Today, more than 120 pulsars have been observed to orbit around planets, main-sequence stars, evolved giants, semi-degenerate stars, white dwarfs and neutron stars. Like the original Hulse–Taylor binary, their clock-like properties allow for precision measurements of their orbital dynamics that can be used to infer stellar properties, probe the physics of binary evolution and test the predictions of General Relativity and alternative theories of gravity.

J. Antoniadis, *Multi-Wavelength Studies of Pulsars and Their Companions*, Springer Theses, DOI 10.1007/978-3-319-09897-5_2

2.2 Timing and Orbits

As briefly mentioned before, much of the interesting science related to radio pulsars comes from the regular monitoring of their rotation.

For any astrophysical source, the time of arrival (TOA) of an emitted signal depends on its (changing) distance from the earth; for pulsars, the signal of interest is the pulse that sweeps the Earth once-per-rotation. Because the rotation is nearly constant, the rotational phase ϕ corresponding to a time of emission t can be approximated by a Taylor expansion:

$$\phi(t) = \phi_0 + \nu(t - t_0) + \frac{1}{2}\dot{\nu}(t - t_0)^2 + \cdots , \tag{2.1}$$

where ϕ_0 and t_0 are arbitrarily chosen reference phase and time. The salient property that enables precision measurements is that the difference between any two times of emission has to be an integer number, $\Delta\phi = N \in \mathbb{Z}$. The TOA differs from t by an amount that depends on propagation delays due to the motion of the Earth, the interstellar medium and the motion of the pulsar:

$$\Delta t = \Delta_{\mathrm{E}\odot} + \Delta_{\mathrm{R}\odot} + \Delta_{\mathrm{S}\odot} - D/f^2 + \Delta_{\mathrm{Binary}}. \tag{2.2}$$

Here, the first three terms account for the Einstein, Roemer and and Shapiro delays of the bodies in the Solar System; the fourth term is the contribution due to the dispersion of the signal from the interstellar medium at an observing frequency f and the fifth term accounts for the binary motion of the pulsar and secular terms due to the system's motion as a hole (Lorimer and Kramer 2005). We shall now focus on the last term adopting the convention of Damour and Deruelle (1986) and Damour and Taylor (1992). A summary of the basic (Keplerian) orbital elements and naming conventions used throughout this thesis can be seen in Fig. 2.1. Obviously, Δ_{Binary} has to be a function of the orbital period, time of ascending node passage, eccentricity and projected semi-major axis ($x \equiv a \sin i/c$):

$$\left\{p^{\mathrm{K}}\right\} = \{P_{\mathrm{b}}, T_0, e_0, \omega_0, x_0\}, \tag{2.3}$$

where all subscripted values refer to an arbitrary epoch. For the detailed description of the binary motion we also need an additional set of parameters that can model any possible deviation from the classical Keplerian motion:

$$\left\{p^{\mathrm{PK}}\right\} = \left\{\dot{P}_{\mathrm{b}}, \gamma, r, s, \delta_\theta, \dot{e}, \dot{x}, \dot{\omega}, \delta_{\mathrm{r}}, A, B, D\right\}. \tag{2.4}$$

Here, $\dot{P}$, $\dot{e}$, $\dot{x}$ and $\dot{\omega}$ can be thought of as the first term of the Taylor expansions for the relevant parameters. Damour and Taylor (1992) showed that the above combination of Keplerian and *post-Kelperian* (PK) terms can correct for deviations (up-to) $O(v^5/c^5)$ weaker than the Newtonian gravitational interaction:

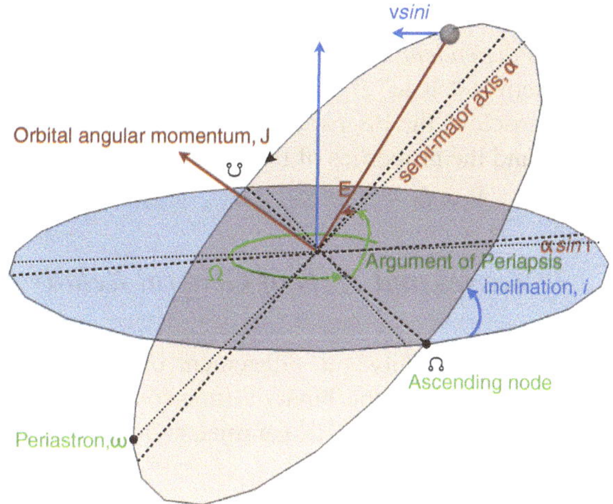

Fig. 2.1 Angles and orientation related to the pulsar orbit

$$\Delta_{\rm R} = D^{-1}\left\{x\sin\omega\,[\cos u - e(1+\delta_{\rm r})] + x[1-e^2(1+\delta_\theta)^2]^{1/2}\cos\omega\sin u\right\}, \tag{2.5}$$

$$\Delta_{\rm E} = D^{-1}\gamma\sin u, \tag{2.6}$$

$$\Delta_{\rm S} = -D^{-1}2r\ln\left\{1 - e\cos u - s\left[\sin\omega\,(\cos u - e) + (1-e^2)^{1/2}\cos\omega\sin u\right]\right\}, \tag{2.7}$$

$$\Delta_{\rm A} = D^{-1}\left\{A[\sin(\omega + A_{\rm e}(u)) + e\sin\omega] + B[\cos(\omega + A_{\rm e}(u)) + e\cos\omega]\right\}, \tag{2.8}$$

$$A_{\rm e}(u) = 2\arctan\left[\left(\frac{1+e}{1-e}\right)^{1/2}\tan\frac{u}{2}\right], \tag{2.9}$$

$$\omega = \omega_0 + \kappa\Delta_{\rm e}(u) \tag{2.10}$$

and

$$u - e\sin u = 2\pi\left[\left(\frac{t-t_0}{P_{\rm b}}\right) - \frac{1}{2}\dot{P}_{\rm b}\left(\frac{t-t_0}{P_{\rm b}}\right)^2 + \cdots\right]. \tag{2.11}$$

Here, D^{-1} can be thought of as a Doppler term due to the secular motion of the center of mass and can be reabsorbed into re-definitions of the other parameters so that one can adopt $D = 1$ for the fitting procedure (Damour and Taylor 1991; Lorimer and Kramer 2005). Similarly, A and B are non-separable and can be neglected by redefining T_0, x, e, $\delta_{\rm r}$ and δ_θ. Equations (2.5)–(2.11) resemble the Roemer, Einstein, Shapiro and aberration delays respectively; δ_θ and $\delta_{\rm r}$ quantify possible relativistic orbital deformations of order $O(v^2/c^2)$ and $\{r, s\}$ parametrize the "range" and "shape" of the Shapiro delay. An important thing to note is that the former timing formula is *theory independent* and describes the orbit in a phenomenological manner. In case the system is "clean" (i.e the orbiting bodies can be approximated by non-rotating point particles), the PK parameters are (in the most general case) functions of their masses and the properties of their internal gravitational field (Will 1993).

2.2.1 Masses and Tests of General Relativity

In General Relativity, the effacement of the internal structure, the gravitational interaction between the binary's (non-rotating) components is only a function of their masses. Hence, the PK parameters become functions of (only) the masses and Keplerian parameters:

$$\dot{\omega} = 3\left(\frac{P_b}{2\pi}\right)^{-5/3}(T_\odot M)^{2/3}\left(1-e^2\right)^{-1}, \tag{2.12}$$

$$\gamma = e\left(\frac{P_b}{2\pi}\right)^{1/3} T_\odot^{2/3} M^{-4/3} m_2\,(m_1+2m_2)\,, \tag{2.13}$$

$$\dot{P}_b = -\frac{192\pi}{5}\left(1+\frac{73}{24}e^2+\frac{37}{96}e^4\right)\left(1-e^2\right)^{-7/2} T_\odot^{-5/3} m_1 m_2 M^{-1/3}, \tag{2.14}$$

$$r = T_\odot m_2 \tag{2.15}$$

$$s = \sin i = x\left(\frac{P_b}{2\pi}\right)^{-2/3} T_\odot^{-1/3} M^{2/3} m_2^{-1} \tag{2.16}$$

and

$$\delta_r = \delta_\theta\,. \tag{2.17}$$

Here, $T_\odot = GM_\odot/c^3$ and $M = m_1 + m_2$ is the total mass of the binary. We note that the former set of equations refers to the intrinsic PK effects that can be extracted from the measured values after taking into account kinematic corrections (i.e. finding $\dot{D}/D$): if these were constant, their contribution would cancel-out directly. Unfortunately, this is normally not the case, since the accelerated motion of the system in the Galaxy results in secular variations that need to be corrected explicitly.

The measurement of *two* post-Keplerian parameters defines a set of functions on the mass-mass plane, the intersection of which yields the masses of the pulsar and the companion. If a *third* PK parameter becomes measurable and General Relativity is correct, the corresponding mass-mass curve should intersect with the previous two at the same point. For the original Hulse–Taylor binary the first PK parameters to be measured were γ and $\dot{\omega}$. Some years after, Weisberg and Taylor (1981) announced the detection of orbital decay which was a strong-field relativistic effect entering the orbital dynamics at the 2.5 ($O(v^5/c^5)$) Post-Newtonian level in GR (Will 1993). The measured value followed the prediction of the quadrupole formula (Eq. 2.13) providing the first radiative test of General Relativity.

Today, the Hulse–Taylor pulsar is outshined by the Double Pulsar for which all the former PK parameters have been measured. Owing to its short orbit (2.45 h) and proximity to the Earth, the kinematic effects can be constrained with high accuracy, yielding 5 distinct tests of General Relativity (including Shapiro delay).

Similarly, all proposed alternative theories of gravity result to unique formulas for the PK parameters. In the following Chapters we focus on two different families of alternative theories:

1. **Scalar-Tensor Gravity**: Scalar-Tensor (ST) theories of gravity are extensions of General Relativity in which gravity is mediated by a spin-2 graviton and a spin-0 scalar partner, ϕ. The motivation for these theories is multi-fold and related, among-else, to Grand Unification attempts and questions concerning Dark Matter, Dark Energy and Inflation.
   In ST-gravity the Strong Equivalence Principle is violated, leading to emission of dipole gravitational radiation that enters the orbital dynamics at the 1.5 Post-Newtonian level (Will 1993). The amount of dipolar waves emitted by the system depends on the difference of the binding energies inside the two bodies. Systems composed of two neutron stars, like the Double Pulsar and the Hulse–Taylor binary are therefore less sensitive to dipolar waves. The predictions of ST gravity can be best tested with "clean" assymetric systems composed of a pulsar and a white dwarf.
2. **Tensor-Vector-Scalar Gravity (TeVeS)**: TeVeS is a relativistic formulation of Modified-Newtonian-Dynamics (MOND) designed to explain galaxy-rotation curves without invoking Dark Matter. The gravitational interaction is mediated by spin–2, spin–1 and spin–0 particles that lead to modifications of all PK parameters. These theories can be tested with both neutron star/neutron star and neutron star/white dwarf binaries.

2.2.2 Special Cases: Circular Orbits

As we discuss further bellow, many millisecond pulsars reside in short-period, circular-orbit binaries. For these systems most PK parameters vanish and only the Shapiro delay and orbital decay can be measured. The former depends strongly on the inclination and therefore can be constrained only for systems viewed nearly edge-on; the latter is sensitive to the orbital period and can be measured only in "relativistic binaries", i.e. systems with short orbital periods.

2.2.3 Special Cases: Mass Ratios and Spectroscopy

For a handful of binary millisecond pulsars, the companion is bright enough for phase-resolved optical spectroscopy. This allows the measurement of its radial velocity which, together with the radial velocity of the pulsar measured with radio-timing, yields the mass ratio of the system. Furthermore, in case the companion is a white dwarf, comparison of its spectrum with model atmospheres yields its mass. Combined, the mass ratio and companion mass yield the mass of the pulsar. This information allows for strong-field radiative tests in relativistic binaries, even if constraints on Shapiro delay are not possible. Similarly, for the Double Pulsar where both neutron stars are visible, the Roemer delays yield a theory-independent mass ratio.

2.3 Recycled Pulsars and Their Formation

With few exceptions, the fastest spinning pulsars known, have white-dwarf or semi-degenerate companions. These systems share remarkable similarities, thought to be the relics of their evolutionary history: They spin down $\sim 10^3$–10^5 times slower than their single counterparts (Lorimer and Kramer 2005) which implies that they have relatively weak magnetic fields of order $\sim 10^8$ G. Furthermore, their orbits are almost perfectly circular and in some cases the mass of their companions is so low that they would not have formed within a Hubble-time if they were single stars.

Today, it is firmly established that these binaries emerge from systems initially formed by a massive, $M > 6\,M_\odot$ star (the progenitor of the pulsar) and a lighter companion. After formation of the neutron star, the system evolves on a timescale determined by the orbital separation and companion mass. For donor star masses above $2.5\,M_\odot$ and short initial periods, the evolution off the main sequence results in engulfment of the neutron star in the donor's envelope. Efficient removal of angular momentum during the common-envelope (CE) phase shrinks the orbit on a very short timescale ($\sim 10^4$ years) (Tauris and van den Heuvel 2003). If the orbital separation is larger than donor's radius during its entire lifespan, mass accretion will initiate when (and if) the donor fills its Roche-lobe. During this period the binary is observed as an X-ray binary. In X-ray binaries, the accretion episode can be long-lasting (10^8–10^{10} years) and thereby allowing the neutron star to accrete sufficient mass (and angular momentum) and spin it up to millisecond periods. Furthermore, the developed tidal torques synchronize the donor on a short timescale resulting in almost perfectly circularized orbits. For what follows, we shall only consider cases with donor star masses $\leq 2.5\,M_\odot$ that ultimately lead to formation of low-mass white dwarfs or semi-degenerate stars. For a recent general review on the evolution of other systems see Tauris and van den Heuvel (2003).

2.3.1 Evolution of the Orbital Separation

During the X-ray binary phase, transfer of angular momentum changes dramatically the orbital dynamics. The orbital angular momentum is given by:

$$J_{\rm orb} = \frac{m_1 m_2}{M}\Omega a^2 \sqrt{1-e^2}, \tag{2.18}$$

where $\Omega = \sqrt{GM/a^3}$ is the angular orbital velocity. We can find the evolution of the orbital separation, by differentiating the above equation:

$$\frac{\dot{a}}{a} = 2\frac{\dot{J}_{\rm orb}}{J_{\rm orb}} - 2\frac{\dot{m}_1}{m_1} - 2\frac{\dot{m}_2}{m_2} + \frac{\dot{m}_1 + \dot{m}_2}{M}. \tag{2.19}$$

Here, the total change in angular momentum can be thought of as the sum of contributions due to gravitational radiation, magnetic braking, spin-orbit couplings and mass-loss from the system (Tauris and van den Heuvel 2003):

$$\frac{\dot{J}_{\rm orb}}{J_{\rm orb}} = \frac{\dot{J}_{\rm GW}}{J_{\rm orb}} + \frac{\dot{J}_{\rm mb}}{J_{\rm orb}} + \frac{\dot{J}_{\rm ls}}{J_{\rm orb}} + \frac{\dot{J}_{\rm ml}}{J_{\rm orb}}, \tag{2.20}$$

where we have neglected changes to the eccentricity, because tidal interactions circularize the orbit on a much sorter timescale. Depending on the initial orbital separation, Roche-lobe overflow (RLO) can initiate when the star is still on the main sequence (Case-A RLO), during the RGB phase (Case-B RLO) or during helium shell-burning (Case-C RLO).

Case-A Roche-Lobe Overflow For short initial separations RLO initiates while the star is still on the main sequence. The evolution of these systems is driven by angular momentum loses due to magnetic braking (MB) and mass ejection from the system.

MB is thought to be the main mechanism responsible for the deceleration of low-mass stars with convective envelopes. In binaries, it operates at the expense of orbital angular momentum due to the tidal torques that tent to synchronize the spin. The details of the MB mechanism are uncertain but it seems that the dependence between the angular momentum loss and the stellar parameters is of the form:

$$\frac{\dot{J}_{\rm mb}}{J_{\rm orb}} \simeq -0.5\times 10^{-28} f_{\rm mb}^{-2} \frac{k^2 R_2^4}{a^5}\frac{GM^3}{m_1 m_2}\,{\rm s}^{-1} \tag{2.21}$$

where k^2 is the gyration radius of the donor and $f_{\rm mb}$ a constant of order unity (Tauris and van den Heuvel 2003).

Additionally, mass loss from the system results in an angular momentum loss rate given by:

$$\frac{\dot{J}_{\rm ml}}{J_{\rm orb}} = \frac{\alpha + \beta q^2 + \delta\gamma(1+q)^2}{1+q}\frac{\dot{m}_2}{m_2} \tag{2.22}$$

where α, β and δ are the fractions of mass lost through a direct wind, mass ejected (uniformly) from the accretor and from a circumbinary coplanar toroid with radius $r = \gamma^2\alpha$.

Systems in this category evolve with decreasing orbital periods and eventually form binaries with a semi-degenerate companion (a.k.a "black widow" systems) or, perhaps, relativistic binaries with white dwarf companions (see next Chapters).

Case-B Roche-Lobe Overflow For larger initial neutron star-donor separations ($P_b \geq 2$ d, Tauris and Savonije 1999), RLO initiates when the star evolves to a sub-giant and starts climbing its Hayashi track on the H-R diagram. The angular momentum loss mechanisms are generally not important and these systems evolve with increasing orbital period and descent to binaries with helium-core white dwarf companions.

For stars on the RGB, the growth of the helium core is directly related to the luminosity which is generated entirely by hydrogen-shell burning. Additionally, during this phase the temperature remains nearly constant and therefore the luminosity is also proportional to the stellar radius ($L = 4\pi\sigma T^4 R^2$) which is equal to the Roche lobe radius. Consequently, the mass of the core is correlated with the orbital period and therefore the *final* mass of the white dwarf is also a function of the *final* orbital period. This theoretical mass-orbital period relation has been studied extensively in the literature (Pylyser and Savonije 1989; Tauris and Savonije 1999, e.g) and seems to follow fairly well the observational data (Fig. 2.2).

An additional relation that can be verified observationally is a positive correlation between the orbital period and the eccentricity arising from tidal perturbations due to the convective envelope of the donor that prohibit perfect circularization (Phinney 1992).

Case C Roche-Lobe Overflow In this case the initial separation is wide and mass transfer initiates when the star fuses its helium layer to carbon. These systems descent to binary pulsars with typical $\sim$0.4 $M_\odot$ white dwarf companions.

2.4 Low-Mass He-Core White Dwarf Companions

Low-mass white dwarf companions accompanying millisecond pulsars are, in principle, very simple objects: they consist of a degenerate helium core, surrounded by a residual hydrogen envelope of size inversely proportional to the core mass (see Chap. 5). Unlike regular white dwarfs, the main source of energy is not the latent heat of the core but residual hydrogen burning in the envelope. This allows them

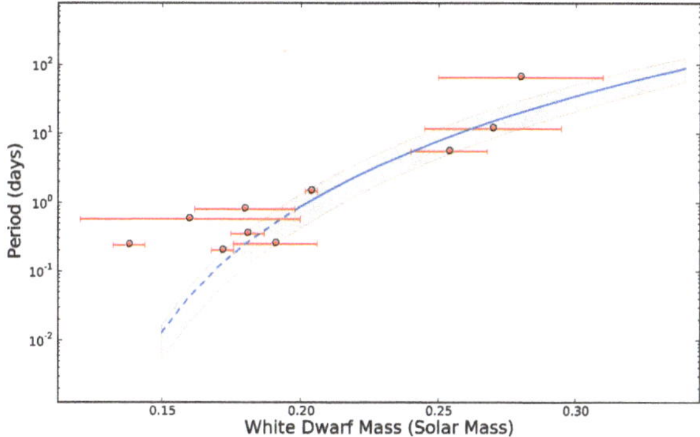

Fig. 2.2 White dwarf mass–orbital period relation based on the detailed binary evolution calculations of Tauris and Savonije (1999). The *shaded area* reflects the underlying modeling uncertainties due to the metallicity of the progenitor, which affects the cessation of mass-transfer through regulation of the magnetic-braking and size of the convective zone. The *dotted line* is an extrapolation of the relation to values below the bifurcation period where, in principle, the correlation does not hold. The *points* depict all currently known systems with determined masses. The plotted periods are inferred from the current orbital-periods and cooling ages of the white-dwarf companions, assuming that the orbit after the Roche-lobe decoupling phase was only affected by gravitational-wave damping

to stay hot and therefore be observable for timescales of several Gyr (van Kerkwijk et al. 2005).

The first low-mass white dwarf detected in the optical was the companion of PSR B0655+66 (Kulkarni 1986). It was immediately recognized that its cooling age can be used as an independent clock that can probe the magnetic field decay of the pulsar. Furthermore as mentioned above, sufficiently hot white dwarfs could be used as tools to infer pulsar masses independently of any post-Keplerian parameters.

In practice, both these uses are somewhat complicated by several issues: First, the calibration of white dwarf cooling ages requires detailed modelling of diffusive and convective processes in their interiors as well as treatment of non-gray effects on their atmospheres (van Kerkwijk et al. 2005). Furthermore, gravitational settling of CNO nuclei towards the core results in runaway hydrogen burning that consumes a large fraction of the envelope and ceases the stable nuclear fusion. It is now established that hydrogen flashes are significant only for white dwarfs more massive than a certain threshold. The latter however depends on a large number of parameters and remains poorly constrained. In the next Chapter we discuss this in more detail.

References

Damour T., Deruelle N., 1986, Ann. Inst. H. Poincare (Physique Theorique), 44, 263
Damour T., Taylor J. H., 1991, ApJ 366, 501
Damour T., Taylor J. H., 1992, Phys. Rev. D, 45, 1840

Giacconi R., Gursky H., Paolini F. R., Rossi B. B., 1962, Physical Review Letters, 9, 439
Hulse R. A., Taylor J. H., 1974, ApJ, 191, L59
Hulse R. A., Taylor J. H., 1975, ApJ, 195, L51
Kulkarni S. R., 1986, ApJ, 306, L85
Lorimer D. R., Kramer M., 2005, Handbook of Pulsar Astronomy. Cambridge University Press
Phinney E. S., 1992, Phil. Trans.:Phys. Sc. & Eng., 341, 39
Pylyser E., Savonije G. J., 1989, A&A, 208, 52
Schreier E., Levinson R., Gursky H., Kellogg E., Tananbaum H., Giacconi R., 1972, ApJ, 172, L79
Tauris T. M., Savonije G. J., 1999, A&A, 350, 928
Tauris T. M., van den Heuvel E. P. J., 2003, ArXiv Astrophysics e-prints
van Kerkwijk M., Bassa C. G., Jacoby B. A., Jonker P. G., 2005, Optical Studies of Companions to Millisecond Pulsars
Weisberg J., Taylor J., 1981, Gen. Relativ. Gravit., 13, 1
Will C. M., 1993, Theory and Experiment in Gravitational Physics

Chapter 3
An Observational Test for Low-Mass Helium-Core White-Dwarf Models

Abstract In this work we present a spectroscopic and photometric analysis of the low-mass white dwarf companion to the pulsar PSR J1909–3744 for which radio timing has yielded a precise mass estimate and an accurate paralactic distance. Based on these we infer the temperature $T_{\rm WD} = 9{,}050 \pm 150\,{\rm K}$, radius, $R_{\rm WD} = 0.0301 \pm 0.015\,{\rm R}_{\odot}$ and surface gravity of the WD, both in a model independent manner ($\log g_{\tau} = 6.77 \pm 0.04$) and using model atmospheres ($\log g = 7.13 \pm 0.15$). We find that , for that range of masses and temperatures, the atmospheric models overestimate the surface gravity by $\sim$5 %, as expected for WDs with fully convective envelopes. Furthermore we show that thick-envelope mass-radius relations reproduce accurately the mass and radius of the WD and they can therefore be used to infer reliable masses for WD companions of other millisecond pulsars.

3.1 Introduction

An interesting group among the degenerate remnants of stars are white dwarfs with such low mass ($\lesssim 0.35\,{\rm M}_{\odot}$) that they must consist mostly of helium. These helium white dwarfs (hereafter, He WD) cannot have formed from single stars, since the required low-mass progenitors have lifetimes well in excess of the age of the Universe. Instead, their formation requires interaction in a binary, in which a $2.5\,{\rm M}_{\odot}$ star looses its hydrogen envelope to a companion as it ascends the red giant branch, before it can ignite helium fusion. Indeed, spectroscopic follow-up confirmed that "low-mass white dwarfs need friends" (Marsh et al. 1995), and their companions range from other white dwarfs (helium and carbon-oxygen) to main sequence stars, sub-dwarfs, and neutron stars (NSs).

Low-mass He WD companions had been expected for many millisecond pulsars in binaries, since these are thought to be spun up in a preceding phase as a low-mass X-ray binary, in which they accrete mass from a low-mass ($\lesssim 1.5\,{\rm M}_{\odot}$) companion. But the first bright counterpart, to the pulsar PSR J1012+5307, led to a surprise: for the age of $\sim$8 Gyr inferred from the pulsar's spin-down rate, the white dwarf was surprisingly hot: $\sim$9,000 K. Alberts et al. (1996) suggested that this might reflect a hydrogen envelope thick enough that it could sustain residual hydrogen burning. The

J. Antoniadis, *Multi-Wavelength Studies of Pulsars and Their Companions*,
Springer Theses, DOI 10.1007/978-3-319-09897-5_3

possibility of very low-mass WDs having sufficiently thick hydrogen envelopes was pointed out by Webbink (1975), and has been confirmed by numerous studies since (e.g. Driebe et al. 1998a; Serenelli et al. 2001). A major uncertainty, however, remains the limiting mass: whether or not a last flash occurs that removes the envelope seems to depend not just on the mass, but also on the amount of CNO present (which in turn depends on metallicity) and the amount of settling and mixing of metals. As a result, estimates of the critical mass range from $\sim$0.17 to $\sim 0.2\,M_{\odot}$. Pulsar binaries may help resolve this question, since they do not just provide good prospects for mass measurements, but also an independent clock (van Kerkwijk et al. 2005).

The companion of PSR J1012+5307 (Nicastro et al. 1995) was followed up spectroscopically by van Kerkwijk et al. (1996) and Callanan et al. (1998), with the intention to use model-atmosphere fits to determine the temperature and surface gravity, combine these with a mass-radius relation to estimate the WD mass, and then infer the NS mass using the mass ratio. The two model-atmosphere fits, however, yielded different values of the surface gravity, subsequently traced to differences in the models. Amusingly, the WD masses were nevertheless the same, since the authors used different mass-radius relations (van Kerkwijk et al. 2005).

Given the above problems, it would be good to test both the model-atmosphere analysis and the mass–radius relation empirically. If we could determine the WD masses with confidence, we could not just determine accurate masses of NSs—and test models for their ultra-dense interiors—but also transform systems with independent post-Keplerian measurements into laboratories for testing strong-field gravity (e.g., PSR J1738sps0333, Freire et al. 2012, see also next chapter). Furthermore, we could characterize with more confidence other binaries, such as the large number of He WDs with white-dwarf companions unveiled in the the SDSS sample (e.g. see Eisenstein et al. 2006; Kilic et al. 2010 and references therein).

Here, we present a test using the He WD companion of PSR J1909−3744, a pulsar for which radio timing has yielded accurate masses as well as an accurate timing parallax (Jacoby et al. 2005). We present our spectroscopy and photometry in Sect. 3.2, and in Sect. 3.3 use them to infer temperature, gravity, radius, and mass ratio, checking the consistency of the mass and radius with the inferred surface gravity, and the consistency of all three with current mass–radius relations. We discuss the ramifications of our work in Sect. 3.4.

3.2 Observations

3.2.1 Spectroscopy

Our dataset consists of fourteen long-slit phase-resolved spectra obtained with the Gemini Multi Object Spectrograph (GMOS) of the Gemini-South telescope over the course of seven nights between April 2004 and July 2005 (see Table 3.1). GMOS is equipped with a mosaic of three 2,048 $\times$ 4,608 CCD arrays which we read out

Table 3.1 Log of observations and radial velocities for PSR J1909−3744 and the comparison star for the Gemini and Keck datasets.

ID	Date	MJD_{bar}	ϕ_{bar}	υ_{WD}	υ_{C}
Gemini					
1	19 April 2004	53114.264729	0.2048	−127 (18)	59.0 (2)
2	19 April 2004	53114.317736	0.2403	−56 (20)	54.1 (2)
3	19 June 2004	53175.097504	0.8789	−221 (18)	52.9 (2)
4	19 June 2004	53175.148732	0.9123	−239 (17)	57.0 (2)
5	17 July 2004	53203.077298	0.1253	−210 (16)	60.9 (2)
6	17 July 2004	53203.129993	0.1596	−130 (13)	67.3 (2)
7	10 Sep 2004	53258.114798	0.0146	−231 (32)	68.6 (2)
8	10 Sep 2004	53258.169397	0.0502	−283 (27)	62.3 (2)
9	11 Sep 2004	53259.033404	0.6135	+66 (24)	68.9 (2)
10	12 May 2005	53502.345332	0.2839	−27 (17)	80.1 (2)
11	12 May 2005	53502.396567	0.3173	+37 (23)	85.3 (2)
12	07 June 2005	53528.395516	0.2389	−51 (17)	81.6 (2)
13	07 June 2005	53528.395516	0.2729	+6 (17)	87.2 (2)
14	11 July 2005	53562.052169	0.2216	−66 (15)	78.5 (2)

binned by two, giving a spatial scale of $0''.146\,\mathrm{pix}^{-1}$. With the 1,200 line mm^{-1} B1200 grism centered at 4,300 Å, we covered 3,500–5,100 Å at 0.4 pix^{-1}.

The slit was oriented at position angle 274°.58 (N through E) to include a bright star 34″ East of the target, which we use for local flux and velocity calibration. To minimize slit losses due to differential atmospheric refraction, we observed with a wide, 1.″5 slit, and guided at a wavelength of 4,300 Å (after acquiring through a g'−band filter). The exposure time for all frames was 3,600 s. The conditions were generally good to photometric and the seeing ranged from 0″.7 to 1″.3, giving a resolution of ∼3 Å (or $\sim 200\,\mathrm{km\,s^{-1}}$) at 4,300 Å.

All science exposures were followed by a flat-field lamp exposure and a Copper-Argon (CuAr) arc exposure for wavelength calibration. Finally, on June 6th 2005 under photometric conditions, we used the same setup, but with a 5″.0 slit to acquire a 60 s exposure of the spectrophotometric flux standard Feige 110 and a 900 s exposure of PSR J1909−3744 and the local reference star.

The data reduction was performed inside the Munich Image Data Analysis System (MIDAS) and follows closely the one presented in detail in a companion paper for PSR J1738+0333 by Antoniadis et al. (2012) (hereafter AVK+12, see also next chapter).

During the reduction, and after converting raw ADUs to electrons using the gain values listed in the GMOS website,[1] we noticed that the column averaged flux in flat fields displayed gaps between chips, suggesting that the amplifiers' gain might

[1] http://www.gemini.edu/sciops/instruments/gmos/?q=node/10477.

be slightly miss-estimated. We also found that the effect was varying from night to night. After accounting for that in all frames (using as reference the middle CCD) and flat-fielding our exposures (as in AVK+12), we subtracted the sky by fitting a second-degree polynomial along the spatial direction to clean regions between the stars and extracted the spectra using a method similar to that of Horne (1986).

We established the dispersion solution by fitting a second-degree polynomial to the identified lines' positions. This gave root-mean-square (rms) residuals of less than $\sim$0.04 Å (or $\leq$1 km s^{-1} at 4,300 Å).

The wide-slit spectra of the comparison and Feige 110 were processed similarly. After correcting all spectra for atmospheric extinction using the average extinction table for La Silla (which should also be reliable for Cero Pachòn) we calculated the flux losses due to the finite size of the slit by comparing the wide-slit spectrum of the comparison with each of the narrow slit spectra. The relation was approximated with a quadratic function of wavelength which was then applied to narrow slit observations.

Finally, we derived the instrumental response of GMOS by comparing our Feige 110 spectrum to its HST/STIS template and smoothly interpolating the ratio. The latter is tabulated at 3 Å and thus before comparing we accounted for the (small) difference in resolution by convolving the template with a Gaussian kernel. We used this to flux calibrate the narrow-slit spectra.

3.2.2 *Photometry*

We analysed all available g'-band acquisition images taken at the beginning of each set of observations before the spectral observations. Following standard practice, we de-biased and flat-fielded the frames, measured the fluxes of the WD and the reference star inside $3''.6$ apertures and subsequently scaled them up to $7''.0$ radii. Using all available 11 measurements, we find that the magnitude difference with reference star is $\Delta g = 4.789(7)$ mag. We found no apparent signs for variability.

We flux-calibrated our measurements using data taken on two photometric nights. First on 2004 Sep 10, a g'-band image was taken of the field containing DMSB 2139−0405, for which SDSS photometry is available. Using this, we infer $g' = 21.87(2)$. Second, on 2005 June 6, images were taken of NGC 4550 and LTT 7379. These yield $g' = 21.88(2)$.

3.3 Results

3.3.1 *Radial Velocities and Orbit*

The radial velocities of the WD and the comparison were inferred via cross-correlation with templates using the method discussed in Bassa et al. (2006). For the comparison star we first identified it as a G6V star by comparison with classification spectra from the on-line atlas of R. O Gray and then used as template the UVESPOP[2] spectrum of HD 140901 tabulated at 2 Å. For the WD companion we used a template DA model atmosphere determined iteratively as in AVK+12.

Each spectrum was fitted for a grid of velocities from -700 to $+700\,\mathrm{km\,s^{-1}}$ with a step size of $5\,\mathrm{km\,s^{-1}}$. At each velocity step we fitted for the normalization and possible variations with wavelength using a 3^{d} polynomial. Best-fit velocities and errors were determined by fitting a parabola to the χ^2 values to within $60\,\mathrm{km\,s^{-1}}$ of minimum.

We accounted for the spectral resolution of the instrument by convolving the templates with a Gaussian with FWHM equal to the seeing, truncated at the slit width. Best fit values had typical reduced χ^2 values of 1.2 and 1.5 for the WD and the comparison respectively. The velocities were transformed to the Solar-system barycenter and corrected for the $-5.2\,\mathrm{km\,s^{-1}}$ velocity of HD 140901.

The radial velocity of the comparison displayed random scatter with rms $\sim$18 $\mathrm{km\,s^{-1}}$, well above the typical $0.2\,\mathrm{km\,s^{-1}}$ formal errors. This large scatter is likely associated with slit-positioning errors and differential atmospheric diffraction. Thus, for further analysis we chose to use velocities relative to the comparison.

We fitted the WD velocities for a circular orbit using a period of $P_{\mathrm{orb}} = 1.533449474590\,\mathrm{days}$ and epoch of ascending node $T_0 = \mathrm{MJD}\,53630.723214894$ from the timing ephemeris of Jacoby et al. (2005). The fit gave a velocity amplitude of $K_{\mathrm{obs}} = 193 \pm 12\,\mathrm{km\,s^{-1}}$ and a systemic velocity (relative to the comparison) of $\Delta\gamma_{\mathrm{obs}} = -115 \pm 6\,\mathrm{km\,s^{-1}}$ with $\chi^2_{\mathrm{red}} = 1.19$ for 12 degrees of freedom. After rejecting one outlier (ID 8 in Table 3.1) we obtain $K_{\mathrm{obs}} = 187 \pm 11\,\mathrm{km\,s^{-1}}$ and $\gamma_{\mathrm{obs}} = -115 \pm 5\,\mathrm{km\,s^{-1}}$ with $\chi^2_{\mathrm{red}} = 1.02$ for 11 degrees of freedom. The solution corresponds to a mass ratio of $K_{\mathrm{WD}}/K_{\mathrm{PSR}} = K_{\mathrm{WD}}P_{\mathrm{b}}/2\pi c x = M_{\mathrm{PSR}}/M_{\mathrm{WD}} = 7.0 \pm 0.5$. For the systemic velocity γ we adopt $\gamma = -73 \pm 30\,\mathrm{km\,s^{-1}}$ which is the value obtained by fitting the raw barycentric WD velocities. The phase-folded velocities and the best-fit orbit are depicted in Fig. 4.2.

3.3.2 Interstellar Extinction

The reddening towards PSR J1909$-$3744 was traced using the Red Clump Stars method (Durant and van Kerkwijk 2006a). We used 30100 stars from the 2MASS catalogue located within $35'.0$ from the companion. The sample was split in seven 0.5 mag-wide stripes covering the range from $K = 10$ to $K = 13.5$. The $J - K$ distribution in each stripe was then fitted with a Gaussian superposed to a power-law function. For our calculations we assumed $K_0 = -1.65$ and $(J-K)_0 = 0.65$ for the intrinsic luminosity and color of the Helium red giants (inferred from low extinction 2MASS fields) and the relations of Schlegel et al. (1998a). The extinction was found to increase smoothly with distance from $A_V = 0.1$ at 100 pc to a maximum of $A_V = 1.15$ at 1.6 kpc. For the parallactic distance of PSR J1909$-$3744 (1.14 kpc)

[2] http://www.sc.eso.org/santiago/uvespop/DATA.

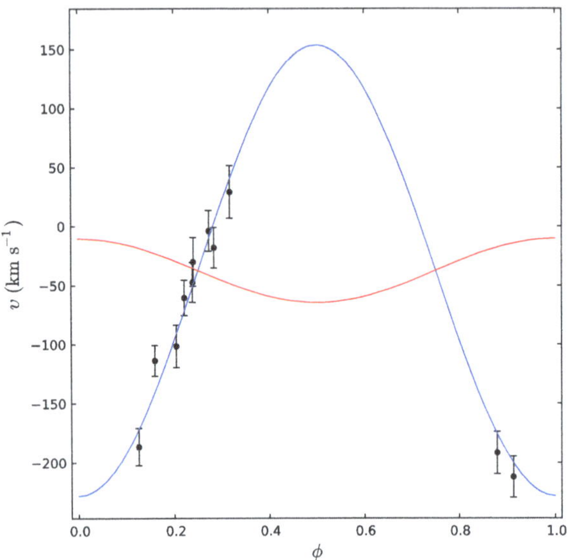

Fig. 3.1 Radial velocity measurements of the companion to PSR J1909−3744 as a function of the orbital phase. Filled *black circles* depict the points used to fit the orbit and the *blue line* the best-fit solution. The *red line* depicts the velocity of the pulsar as inferred from radio timing. All velocities are relative to the comparison star, but corrected for its estimated $-73\,\mathrm{km\,s^{-1}}$ barycentric radial velocity. All *error bars* represent 1σ uncertainties

we measure $A_V = 0.98 \pm 0.06$, hence for $g-$band one infers: $A_g = 1.065 A_V = 1.04{\pm}0.06$. The extinction-corrected apparent magnitude implies a distance modulus of $(m - M)_0 = 10.28 \pm 0.08$ mag and thus $M_g = 10.66 \pm 0.10$.

3.3.3 Spectral Fit

In Fig.3.2 we show the average of the 14 individual Gemini spectra shifted to zero velocity. The spectrum shows deep Balmer lines, from Hβ down to H13, typical of a DA WD with low surface gravity.

The spectrum was fitted with a set of model atmospheres generated by one of us (DK). The atmospheres are a recent update of the ones presented by Koester (2008) that incorporate the improved treatment of pressure broadening of the absorption lines by Tremblay and Bergeron (2009). We scanned a grid of temperatures from $T_{\rm eff} = 6{,}000$ to 20,000 K in steps of 250 K and surface gravities from $\log g = 6.00$ to 8.00 with a step of 0.25 dex. At each point of the grid we fitted for the normalization and possible variations with wavelength using a third degree polynomial (this gave the best fit to higher Balmer lines that are most sensitive to gravity). As above, we accounted for the spectral resolution of the instrument by convolving the templates with a truncated gaussian. For the fit we excluded a small spectral range from 4,500 to 4,700 Å (absent of Balmer lines), were the flux calibration seemed to be imperfect. This region coincides with some features seen in the flat fields likely associated with the holographic grating.

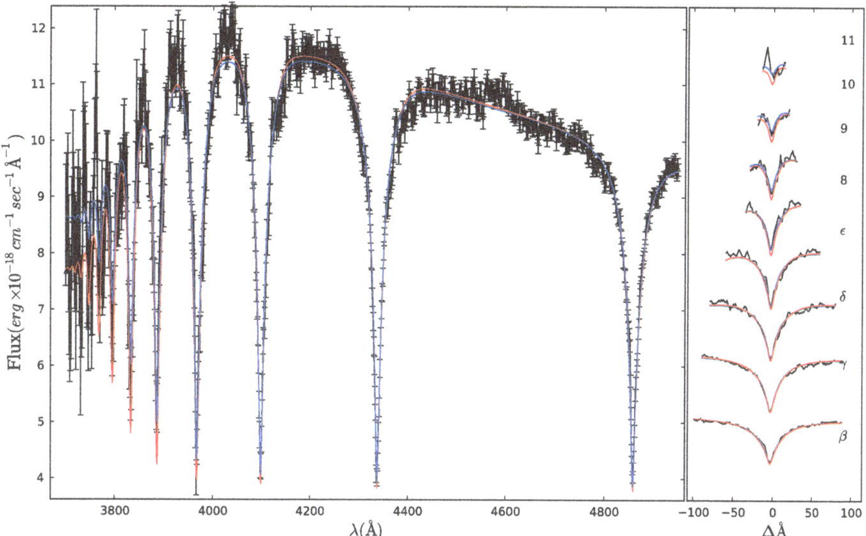

Fig. 3.2 Average spectrum of the WD companion to PSR J1909−3744 created by the coherent addition of 14 individual spectra shifted to zero velocity (see text for details). *Left* The *blue line* depicts the best-fit model-atmosphere corresponding to $T_{\rm eff}$ = 9,100 K and $\log g$ = 7.25. The *red line* is a model-atmosphere corresponding to $T_{\rm eff}$ = 9100 and $\log g$ = 6.77 inferred from photometry and timing. *Right* Details of the *spectral lines*

Best-fit values and their formal uncertainties were determined by fitting a parabola to the χ^2 surface. Using the choices above we find $T_{\rm eff} = 9050 \pm 35$ K and $\log g = 7.13 \pm 0.07$ with $\chi^2 \simeq 1.4$ for 798 degrees of freedom. However, experimentation with our choices for the fit showed that these uncertainties are clearly underestimated and the spectrum is likely polluted by systematics: First, changing the degree of the polynomial for the normalization from 1st to 4th degree results in changes of $\Delta T_{\rm eff} \sim 150$ K and $\Delta(\log g) \sim 0.1$ dex. Second, for the temperature, a fit to individual spectra (keeping $\log g$ fixed to 7.25) yielded an average of $T_{\rm eff}$ = 9,021 ± 30 with an rms scatter of ∼150 K. On the contrary, small changes to the resolution did not affect the results more than 1σ and a fit by one of us (DK) using a different technique yielded T = 9,130 ± 50 K and $\log g = 7.20 \pm 0.12$ that are consistent to the values listed above. Given these results we adopt $T_{\rm eff}$ = 9,050 ± 150 K and $\log g = 7.13 \pm 0.15$ dex with the uncertainties being a conservative estimate of the systematics.

3.3.4 Radius and Surface Gravity

The spectral and photometric observations make it possible to calculate the radius and surface gravity of the WD independently of any modelling assumptions. Convolving the best-fit model ($T_{\rm eff}$ = 9,050 K, $\log g$ = 7.13) with the SDSS g'-band filter curve yields $M_g = 3.09 \pm 0.02$ for a $1\,R_\odot$ object. Combined with the parallactic-distance estimate, $d = 1.14 \pm 0.04\,\text{kpc}$, this implies $R = 0.0306 \pm 0.015\,R_\odot$. Hence, combined with the mass from timing, $M = 0.2038 \pm 0.0022\,M_\odot$, one infers $\log g_\tau = 6.77 \pm 0.04$.

3.4 Ramifications

3.4.1 A Test of the Atmospheric Models

Table 3.2 shows the main observational quantities derived in this chapter. The accurately constrained mass and radius, for the first time allow an independent determination of the WD surface gravity, $\log g_\tau = 6.77 \pm 0.04$. The latter seems to be lower than the measured value $\log g = 7.13 \pm 0.15$. While the difference ($\log g / \log g_\tau = 1.05 \pm 0.02$) is 2σ consistent with zero, the trend resembles that found for higher mass white dwarfs with temperatures $T_{\text{eff}} \leq 12{,}000\,\text{K}$ (Tremblay et al. 2011). For these stars the inconsistency is attributed to the incomplete treatment of convective energy transfer within the framework of the mixing-length theory employed in 1D model atmospheres. For low-mass WDs, one generally expects the same trend, although for somewhat cooler objects, given that higher temperature low-mass WDs are purely radiative (see Chap. 5 for details). Independently of the envelope size, the companion of PSR J1909−3744 is convective and a difference of the same magnitude is expected.

3.4.2 3D Velocity: A Pulsar Coming from the Galactic Center

Combining our analysis in Sect. 3.3.1 with the estimates for the proper motion and parallax of Jacoby et al. (2005), makes it possible to calculate the 3D velocity of PSR J1909−3744: The two components of the transverse velocity are $v_\alpha = \mu_\alpha d = -50.61 \pm 0.01\,\text{km}\,\text{s}^{-1}$ and $v_\delta = \mu_\delta d = -192.32 \pm 0.01\,\text{km}\,\text{s}^{-1}$. Hence the 3D velocity of the system is $|v| = 212 \pm 11\,\text{km}\,\text{s}^{-1}$. This information, allows us to trace the Galactic path of the system back in time. Using the potential of Kenyon et al. (2008) we find that PSR J1909−3744 has a peculiar orbit that intersects the

Table 3.2 Parameters of the He WD companion to PSR J1909−3744, derived throughout this Chapter

Derived parameters	
Temperature (K, spectroscopy)	9,050 (150)
Radius ($R_\odot$)	0.306 (15)
Surface gravity ($\log g$, spectroscopy)	7.13 (15)
Surface gravity ($\log g$, mass+radius)	6.77 (4)

Galactic center every ~100 Myr. We reproduced similar results using the potentials of Kuijken and Gilmore (1989) and Paczynski (1990). Interestingly, we find that the former result is not particularly sensitive to the radial velocity component, γ.

3.4.3 *Comparison of Atmospheric Properties and Mass Radius Relations*

In this section we compare the observational constrains on mass and radius with the predictions of various mass-radius relations found in the literature. Models with thick hydrogen envelopes (Panei et al. 2000; Serenelli et al. 2002) yield a mass of $M_{\rm WD} \sim 0.2\,{\rm M_{\odot}}$ and a radius of $R_{\rm WD} \sim 0.03\,{\rm R_{\odot}}$, both consistent with the independent constraints. On the opposite, relations with thin hydrogen atmospheres (Panei et al. 2000, 2007; Serenelli et al. 2001) predict somewhat smaller masses, e.g. the one of Panei et al. (2000) yields $M_{\rm WD} = 0.172 \pm 0.005\,{\rm M_{\odot}}$, $R_{\rm WD} = 0.028 \pm 0.001\,{\rm R_{\odot}}$. The latter evidence for a thick envelope undergoing residual hydrogen burning is further supported by the pulsar age. Although the characteristic age, $\tau_{\rm c} = 16 \pm 3\,{\rm Gyr}$, is too large to be considered real, it is a strong indicator of a large true age. All thin-envelope models yield an age of ~0.6 Gyr which is too small. On the other thick envelope relations give $\tau_{\rm c} \sim 5\,{\rm Gyr}$ which seems more reasonable.

Interestingly, all thick-envelope mass-radius relations consistent with our observations show a relatively large diversity on input physics. Specifically, PSR J1909−3744 seems to be insensitive to the adopted metallicity and treatment of convection. Therefore observations of lower-mass WDs are necessary to account for the remaining uncertainties.

3.5 Conclusions

We have presented spectroscopic and photometric observations of the companion to PSR J1909−3744. These, for the first time allow a model-independent inference of the radius and surface gravity of a low mass WD. Our comparison with model atmospheres shows that, for that range of masses, the model atmospheres overestimate the surface gravity by ~4 %. This discrepancy is similar to the "high $\log g$" problem found for higher-mass WDs and can be explained by the incomplete treatment of convection in 1D atmospheric models. Hence, this systematic should be absent in lower mass WDs with purely radiative atmospheres (e.g. the ones presented in the next Chapters).

The comparison with cooling tracks shows clear evidence for a thick envelope which is also consistent with the spin-down age of the pulsar. This is generally not predicted by recent calculations that incorporate a detailed treatment of gravitational settling and other diffusive processes. These models yield thin envelopes above a threshold mass of $\sim 0.17 - 0.18\,\mathrm{M}_{\odot}$ as a result of extensive hydrogen flashes due to gravitational settling of CNO nuclei during the pre-WD phase. A possible explanation for the discrepancy could be that the progenitor star had a lower metallicity. An alternative explanation could be that the rotation of the WD plays a significant role in the mixing of elements (Panei et al. 2007). In Chap. 5, I present a new set of WD models which take into account the constraints found here.

References

Alberts F., Savonije G. J., van den Heuvel E. P. J., Pols O. R., 1996, Nature, 380, 676
Antoniadis J., van Kerkwijk M. H., Koester D., Freire P. C. C., Wex N., Tauris T. M., Kramer M., Bassa C. G., 2012, MNRAS, 423, 3316
Bassa C. G., van Kerkwijk M. H., Koester D., Verbunt F., 2006, A&A, 456, 295
Callanan P. J., Garnavich P. M., Koester D., 1998, MNRAS, 298, 207
Driebe T., Schoenberner D., Bloecker T., Herwig F., 1998a, A&A, 339, 123
Durant M., van Kerkwijk M. H., 2006a, ApJ, 650, 1070
Eisenstein D. J., Liebert J., Harris H. C., Kleinman S. J., Nitta A., Silvestri N., Anderson S. A., Barentine J. C., Brewington H. J., Brinkmann J., Harvanek M., Krzesiński J., Neilsen Jr. E. H., Long D., Schneider D. P., Snedden S. A., 2006, ApJS, 167, 40
Freire P. C. C., Wex N., Esposito-Farèse G., Verbiest J. P. W., Bailes M., Jacoby B. A., Kramer M., Stairs I. H., Antoniadis J., Janssen G. H., 2012, MNRAS, 423, 3328
Horne K., 1986, PASP, 98, 609
Jacoby B. A., Hotan A., Bailes M., Ord S., Kuklarni S. R., 2005, ApJ, 629, L113
Kenyon S. J., Bromley B. C., Geller M. J., Brown W. R., 2008, ApJ, 680, 312
Kilic M., Brown W. R., Allende Prieto C., Kenyon S. J., Panei J. A., 2010, ApJ, 716, 122
Koester D., 2008, arXiv: 0812.0482
Kuijken K., Gilmore G., 1989, MNRAS, 239, 605
Marsh T. R., Dhillon V. S., Duck S. R., 1995, MNRAS, 275, 828
Nicastro L., Lyne A. G., Lorimer D. R., Harrison P. A., Bailes M., Skidmore B. D., 1995, MNRAS, 273, L68
Paczynski B., 1990, ApJ, 348, 485
Panei J. A., Althaus L. G., Benvenuto O. G., 2000, A&A, 353, 970
Panei J. A., Althaus L. G., Chen X., Han Z., 2007, MNRAS, 382, 779
Schlegel D. J., Finkbeiner D. P., Davis M., 1998a, ApJ, 500, 525
Serenelli A. M., Althaus L. G., Rohrmann R. D., Benvenuto O. G., 2001, MNRAS, 325, 607
Serenelli A. M., Althaus L. G., Rohrmann R. D., Benvenuto O. G., 2002, MNRAS, 337, 1091
Tremblay P.-E., Bergeron P., 2009, ApJ, 696, 1755
Tremblay P., Ludwig H., Steffen M., Bergeron P., Freytag B., 2011, ArXiv e-prints
van Kerkwijk M., Bassa C. G., Jacoby B. A., Jonker P. G., 2005, Optical Studies of Companions to Millisecond Pulsars
van Kerkwijk M. H., Bergeron P., Kulkarni S. R., 1996, ApJ, 467, L89
Webbink R. F., 1975, MNRAS, 171, 555

Chapter 4
The Relativistic Binary PSR J1738+0333

Abstract PSR J1738+0333 is one of the four millisecond pulsars known to be orbited by a white dwarf companion bright enough for optical spectroscopy. Of these, it has the shortest orbital period, making it especially interesting for a range of astrophysical and gravity related questions. We present a spectroscopic and photometric study of the white dwarf companion and infer its radial velocity curve, effective temperature, surface gravity and luminosity. We find that the white dwarf has properties consistent with those of low-mass white dwarfs with thick hydrogen envelopes, and use the corresponding mass-radius relation to infer its mass; $M_{\rm WD} = 0.181^{+0.007}_{-0.005}\,{\rm M}_{\odot}$. Combined with the mass ratio $q = 8.1 \pm 0.2$ inferred from the radial velocities and the precise pulsar timing ephemeris, the neutron star mass is constrained to $M_{\rm PSR} = 1.47^{+0.07}_{-0.06}\,{\rm M}_{\odot}$. Contrary to expectations, the latter is only slightly above the Chandrasekhar limit. We find that, even if the birth mass of the neutron star was only $1.20\,{\rm M}_{\odot}$, more than 60 % of the matter that left the surface of the white dwarf progenitor escaped the system. The accurate determination of the component masses transforms this system in a laboratory for fundamental physics by constraining the orbital decay predicted by general relativity. Currently, the agreement is within 1σ of the observed decay. Further radio timing observations will allow precise tests of white dwarf models, assuming the validity of general relativity.

4.1 Introduction

Millisecond pulsars (MSPs) are extreme in many ways. Their interior consists of the densest form of observable matter known and they can spin at least as fast as 716 times per second (Hessels et al. 2006). Hence, they offer a rich laboratory for a wide range of physical inquiry: Mass measurements provide direct comparison to quantum chromodynamics' predictions for the state of ultra-dense matter (Lattimer and Prakash 2004; Demorest et al. 2010) and studies of their orbits in binaries have provided the first confirmation for gravitational wave emission and the most stringent

---

Based on the following publications: http://adsabs.harvard.edu/abs/2012MNRAS.423.3316A, http://adsabs.harvard.edu/abs/2012MNRAS.423.3328F.

J. Antoniadis, *Multi-Wavelength Studies of Pulsars and Their Companions*,
Springer Theses, DOI 10.1007/978-3-319-09897-5_4

strong-field tests of general relativity (Taylor and Weisberg 1982; Weisberg et al. 2010; Kramer et al. 2006).

Most of the fastest spinning Galactic-disk pulsars are paired with low mass helium-core WDs [hereafter LMWDs, for recent reviews see (Lorimer 2008; Tauris 2011)], and their fast spins and weak magnetic fields are thought to be the product of mass transfer from the progenitor of the WD, a process also known as recycling. As the progenitor star evolves, it fills its Roche lobe and loses its envelope, either while on the main sequence (for sufficiently short initial periods), or when moving up the red-giant track (Webbink et al. 1983). The mass transfer rate is a strongly increasing function of the initial orbital period and donor mass (Tauris and Savonije 1999), and is expected to be at a stable, sub-Eddington rate ($\lesssim 10^{-8}\,M_{\odot}\,yr^{-1}$) for light companions in relatively tight orbits. The final result of such long-term (nuclear timescale) mass transfer is a highly circular (due to fast tidal dissipation in the secondary) close binary consisting of a fast spinning MSP and a low mass, helium-core WD.

These systems are important for several reasons. First, it is these binaries that allow one to probe certain aspects of the radiative properties of gravity that are poorly constrained by the relativistic effects seen in double neutron stars, like the Hulse–Taylor or the double pulsar. For example, in a wide range of theories, the rate of gravitational wave emission is driven by a leading dipolar term that depends crucially on the difference in gravitational binding energies between the binary members. Hence, if accurate component masses can be determined, one can directly confront the predictions of different gravity theories in terms of dipolar radiation with observations.

Second, measuring their masses provides access to the accretion process and evolution of these systems as well as the formation of MSPs, the only neutron stars with secure precise masses significantly above the Chandrasekhar limit (Freire et al. 2011; Demorest et al. 2010). In addition, observational constraints on the upper mass limit of stable neutron stars, constrains the equation of state for super-dense matter.

Unfortunately, precise MSP and companion masses can be determined from timing in exceptional cases only: either when the orbit is (unexpectedly) eccentric, allowing for a measurement of the rate of advance of periastron (Freire et al. 2011), or if the system has an orbit seen edge on (Kaspi et al. 1994; Jacoby et al. 2005; Demorest et al. 2010) which allows for a measurement of pulse time-of-arrival (TOA) delays due to the curvature of space-time around the companion (Shapiro delay, Shapiro 1964).

Fortunately, another method exists that relies on combined optical and radio timing observations (van Kerkwijk et al. 1996; Callanan et al. 1998). If the WD companion is bright enough for detailed spectroscopy, a comparison of its spectrum with model atmospheres yields its effective temperature and surface gravity. These can then be compared with a mass–radius relation for LMWDs to yield its mass. Combining the radial velocity for the white dwarf with the pulsar timing measurements yields the mass ratio and therefore the mass of the pulsar.

In the previous chapter, we test this method on PSR J1909−3744, for which the masses are precisely known from timing. We find it reliable and are confident to apply it also to other similar systems. In this chapter we report on the application of this method to PSR J1738+0333, a pulsar-LMWD binary, discovered in a Parkes survey (Jacoby et al. 2007). Because of its short orbital period of ~8.5 h and for the reasons mentioned above, the system is of particular interest for radiative tests of gravity. Furthermore it provides a valuable input for binary evolution theory since it lies in a regime where nuclear-driven evolution had most likely been overtaken by magnetic-braking and gravitational radiation (Phinney and Kulkarni 1994) The text is organized as follows: Sect. 4.2 starts with presenting results from radio timing, necessary for calculations throughout the rest of this chapter. These are described in detail in a companion work led by (Freire et al. 2012, Paper II from know on). We then describe the spectroscopic and photometric observations and in Sect. 4.3 we present our results. We discuss our findings and comment on the evolution of the system and its importance for gravity tests in Sect. 4.4. Finally, in Sect. 4.5 we summarize our results and in Sect. 4.6 the conclusions of Paper II.

4.2 Observations

4.2.1 Radio

PSR J1738+0333 was discovered in a 20-cm high Galactic latitude survey in 2001 (Jacoby 2005), carried out with the multi-beam receiver of the Parkes Telescope. The pulsar has a spin period of 5.85 ms and orbits a low-mass helium-core WD companion in a 8.5 h orbit. Since 2003 it has been regularly timed with the 305 m Arecibo Telescope, leading to ~17,000 times of arrival with typically 3 μs uncertainties. The corresponding timing solution provides measurements of the system's parallax and proper motion, and a significant detection of the intrinsic orbital period derivative (see Paper II for details). In Table 4.2 we list the measured spin, Keplerian and astrometric parameters of the system.

The spin period derivative is that of a typical low-surface magnetic field pulsar ($B_0 = 3.7 \times 10^8$ G), and the characteristic age ($\equiv P/2\dot{P}$) after subtracting the kinematic effects (Paper II) is 4.1 Gyr. The parallax measurement corresponds to a distance of $d = 1.47 \pm 0.10$ kpc. The system's proper motion combined with the parallax implies transverse velocities of $v_\alpha = d\mu_\alpha = 49\,\mathrm{km\,s^{-1}}$ and $v_\delta = d\mu_\delta = 36\,\mathrm{km\,s^{-1}}$ in α and δ respectively. In Sect. 4.3 we combine these values with the systemic radial velocity, γ, to derive the 3D spatial velocity and calculate the Galactic orbit of the binary. The estimate for the orbital eccentricity is one of the lowest observed in any binary system: When Shapiro delay is accounted for in the solution (Paper II), the apparent eccentricity diminishes to $e = (3.5 \pm 1.1) \times 10^{-7}$. We discuss the implication of this for evolutionary scenarios in Sect. 4.4.

4.2.2 Optical

Our main data set consists of eighteen long-slit phase resolved spectra of PSR J1738+0333, obtained with the Gemini South telescope at Cerro Pachón on ten different nights between April and June 2006 (see Table 4.1). For our observations we used the Gemini Multi-Object Spectrograph (GMOS-S). The GMOS detector consists of three 2,048 × 4,608 EEV CCDs, each of which was read-out at 2 × 2 binning by a different amplifier, giving a scale of 0.″14 per binned pixel in the spatial direction, and, with the 1,200 lines per mm B1200 grism, 0.4 Å per binned pixel in the dispersion direction. We chose a relatively wide, 1.″5 slit, to minimize atmospheric dispersion losses (see below). This meant that the resolution was set by the seeing, at ∼3 Å, or ∼200 km s^{-1} at 4,300 Å. In order to cover the higher Balmer lines, we centred the grating at 4,300 Å, for a wavelength coverage from 3,500 to 5,100 Å.

All exposures had integration times of 3,720 s and were followed by an internal flat-field exposure and a Copper-Argon (CuAr) exposure for wavelength calibration. The slit was oriented to include a bright comparison star located 25.″2 at position angle 127.°57 (north through east) of the WD (see Fig. 4.1). We use this star as a local velocity and flux standard (since GMOS-S does not have an atmospheric dispersion corrector, slit losses vary with offset from the parallactic angle).

The conditions during the observations were mostly good to photometric, but some exposures were taken through thin cirrus. The seeing ranged from 0.″6 to 1.″2. For flux calibration, we acquired additional frames of the comparison star and the spectro-photometric standard EG 274 through a 5.″0 slit on the night of 2006 April 27 (which was photometric and had 0.″8 seeing). Furthermore, for absolute velocity calibration, we observed the radial velocity standard WD 1743−132 on 2006 June 19.

The data were reduced using standard and custom routines inside the Munich Image and Data Analysis System (MIDAS). First, the bias level of each exposure was removed using average values from the overscan region. Subsequently, we corrected the raw counts on the red and middle chips for the small, few percent variations in gain (see vK+12 for details on the method), that affected several sets of exposures (but fortunately not those of the night the flux calibrator was taken). Finally, the frames were corrected for small-scale sensitivity variations using normalised lamp exposures, where the normalisation was done both along each wavelength position as well as along each spatial position. These normalisation steps were required since the lamp spectra showed rather sharp bumps in the dispersion direction whose position and shape was different from bumps seen in target spectra, and also varied between sets of spectra (possibly because the holographic grating was not illuminated exactly identically between the different exposures), while in the spatial direction they showed striations due to irregularities in the slit.

For sky subtraction, we selected a 100″ region centred on the WD, but excluding 5″ spots around it and the comparison star. Each column in the spatial direction was fitted with a second degree polynomial and the interpolated sky contributions at the positions of the WD and the comparison were removed.

Table 4.1 Log of observations and radial velocity measurements

Date	$MJD_{mid,bar}$(1)	ϕ (2)	v_R ($km\,s^{-1}$) (3)	v_{WD} ($km\,s^{-1}$) (4)	Δv ($km\,s^{-1}$) (5)	$\Delta B'$ (6)
Gemini,GMOS-S						
2006 Apr 27	53852.310219	0.0250	$+56.3 \pm 0.8$	-209 ± 27	-265 ± 27	2.91 ± 0.06
	53852.366314	0.1831	$+49.1 \pm 0.8$	-143 ± 26	-192 ± 26	2.88 ± 0.06
2006 Apr 28	53853.295453	0.8019	$+53.7 \pm 0.5$	-100 ± 14	-154 ± 14	2.88 ± 0.04
	53853.350638	0.9575	$+68.8 \pm 0.6$	-185 ± 15	-254 ± 15	2.88 ± 0.04
2006 May 07	53862.333037	0.2749	$+40.1 \pm 0.5$	-35 ± 13	-75 ± 13	2.83 ± 0.03
	53862.391933	0.4409	$+84.9 \pm 0.6$	$+162 \pm 21$	$+77 \pm 21$	2.87 ± 0.04
2006 May 26	53881.198674	0.4489	$+60.8 \pm 1.1$	$+121 \pm 37$	$+60 \pm 37$	3.03 ± 0.11
	53881.252933	0.6018	$+67.0 \pm 0.9$	$+33 \pm 32$	-34 ± 32	2.97 ± 0.09
2006 May 27	53882.352291	0.7005	$+43.1 \pm 0.5$	-6 ± 15	-49 ± 15	2.85 ± 0.04
2006 May 28	53883.296760	0.3625	$+54.9 \pm 0.6$	$+28 \pm 16$	-27 ± 16	2.86 ± 0.03
	53883.350144	0.5130	$+53.9 \pm 0.6$	$+134 \pm 17$	-189 ± 17	2.85 ± 0.03
2006 Jun 19	53905.174549	0.0264	$+41.5 \pm 0.5$	-226 ± 12	$+80 \pm 12$	2.85 ± 0.04
2006 Jun 23	53909.170100	0.2881	$+39.9 \pm 0.7$	-5 ± 14	-45 ± 14	2.88 ± 0.03
	53909.210618	0.4023	$+95.6 \pm 1.1$	-103 ± 36	-45 ± 36	2.84 ± 0.04
2006 Jun 26	53912.156147	0.7045	$+60.4 \pm 0.5$	-7 ± 14	-67 ± 14	2.85 ± 0.04
	53912.209838	0.8558	$+60.7 \pm 0.6$	-136 ± 15	-197 ± 15	2.88 ± 0.04
2006 Jun 27	53913.120000	0.4212	$+42.5 \pm 0.5$	$+79 \pm 12$	$+37 \pm 12$	2.84 ± 0.04
	53913.176660	0.5809	$+43.6 \pm 0.5$	$+106 \pm 13$	$+62 \pm 13$	2.82 ± 0.04
Keck, LRIS						
2008 Aug 04	54682.377697	0.6224	50 ± 1, $+61 \pm 5$	-2 ± 9	-52 ± 9	3.01 ± 0.05

Notes 1 refers to the barycentric mid-exposure time. (2) is the orbital phase using the ephemeris in Table 4.2. (3) is the comparison's velocity in respect to the solar system barycenter and (4) the raw barycentric velocities of PSR J1738+0333. (5) is the differential velocity used to determine the orbit in Sect. 4.3. Finally, (6) are the differential spectrophotometric magnitudes in B' (equal to B, but limited to the wavelength range covered by our spectra; see Sect. 4.2.2). Here, the errors are the quadratic sum of the photometric uncertainties of the WD and the comparison. For LRIS, two velocities are listed for the comparison star, for the blue and red arm, respectively. For the white dwarf, the velocity is for the blue arm (see text)

Optimally weighted spectra and their uncertainties were extracted using a method similar to that of Horne (1986). The extraction was done separately in each chip and the spectra were merged after flux calibration.

The dispersion solution was established using the CuAr spectra taken after each exposure. First, the 1D lamp spectrum was extracted by averaging the signal over the spatial direction in areas of the chip that coincided with each star. Then the lines' positions were measured and identified and the dispersion relation was approximated with a 3rd degree polynomial that gave root-mean-square residuals of less than 0.04 Å for typically 18 lines.

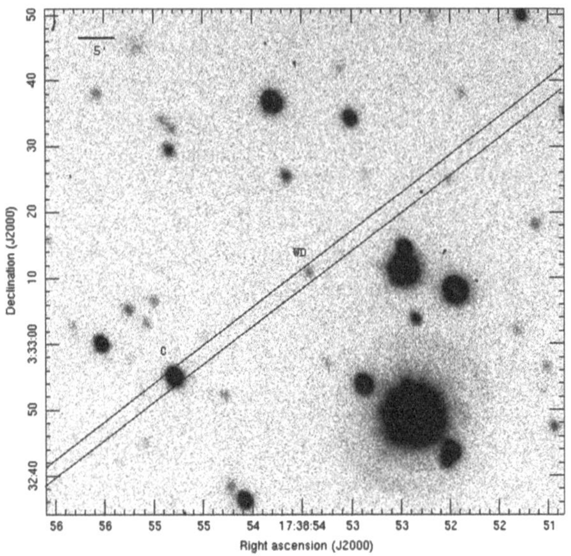

Fig. 4.1 Finding chart for PSR J1738+0333 (using the SOAR *V* image). Indicated are the *white-dwarf counterpart*, the slit orientation used, and the comparison star that was included in the slit

The wide-slit spectra of EG 274 and the comparison star were extracted with the same procedure and used to calibrate the narrow-slit exposures. Initially, all wide and narrow slit data were corrected for atmospheric extinction using the average extinction table for La Silla (which should be a good approximation to that of Cerro Pachón). Then, we calculated the wavelength-dependent flux losses due to the finite size of the slit by comparing the wide-slit spectrum of the comparison with each of the narrow-slit spectra. The relation was analytically approximated with a quadratic function of wavelength that was then applied to the narrow slit observations. Finally, the GMOS instrumental response was calculated by dividing the spectrum of EG 274 with a synthetic template and smoothly interpolating the ratio. The template was created by normalizing an appropriate DA model atmosphere to the catalogued flux ($V = 11.03$, Zwitter et al. 2004, see Sect. 4.5 for more details on the model atmospheres used in this chapter). Prior to comparison, we smoothed the template with a Gaussian kernel to match the resolution of the observed spectrum and excluded the cores of the Balmer lines.

Given the possible issues with the detector gain and the flat fielding, both of which could affect the flux calibration, we obtained an additional smaller set of spectra of the WD companion, the comparison star and the spectro-photometric standard Feige 110 using the two-armed Low Resolution Imaging Spectrometer (LRIS Oke et al. 1995) of the Keck telescope on the night of 2008 August 3 (Table 4.1). During the night the sky was photometric and the seeing was $\sim 0.''8$.

For the observations we used the atmospheric dispersion corrector and both narrow, $0.''7$, and wide, $8.''7$ slits. The light was split with a dichroic at 6800 Å and directed on the two arms of LRIS (blue and red arm hereafter). On the blue arm we used a 600 lines mm^{-1} grism, blazed at 4,000 Å, that covers 3,100–5,600 Å with a resolution of $\Delta\lambda = 3.2$ Å or $\Delta v = 220\,\mathrm{km\,s^{-1}}$. On the red arm we used the 1,200 lines mm^{-1} grating, blazed at 8,000 Å, that covers 7,600–8,900 Å at $\Delta\lambda = 2.1$ Å or $\Delta v = 75$ km s^{-1}. The blue-side detector is a mosaic of two Marconi CCDs with 4,096 × 4,096 pixels 15 μm on the side, which we read out binned by two in the dispersion direction. The red-side detector is a Tektronic CCD with 2,048 × 2,048 pixels 24 μm on the side, which we read-out unbinned.

The spectra were extracted and calibrated as above. Here, on the blue arm we replaced the poorly exposed part of the flat fields shortward of 4,000 Å with unity and normalized the rest using a third degree polynomial. On the red side we normalized the flat field using a bi-linear fit. Wavelength calibration was done using arc spectra and sky lines. On the blue arm we used the well exposed arc frames taken at the beginning of the night to establish an overall solution that had rms residuals of 0.16 Å for 22 lines fitted with a third-degree polynomial and then calculated offsets using the less well-exposed arc frames taken throughout the night. For the red arm we used the well exposed arc-frames taken interspersed with the science exposures. Here, we corrected for offsets by shifting the bright OH and O_2 lines at 8344.602, 8430.174 and 8827.096 Å to laboratory values. Flux calibration was again done as above; we found that the solution was consistent with that obtained from Gemini (see also below).

4.2.3 Photometry

On the night of 2008 February 28, images of the field containing PSR J1738+0333 were acquired for us with the 4.1 m Southern Astrophysical Research Telescope (SOAR) at Cerro Pachón, Chile, using the Goodman High Throughput Spectrograph (Clemens et al. 2004), with its Fairchild 4,096 × 4,096 CCD and B and V filters (with throughputs on the Kron-Cousins photometric system). The instrument has a plate scale of $0.''15$ pix^{-1} and a usable field of view of $5.'0$. During the run, the sky was photometric and the seeing as determined from the images was $\sim 1.''8$. Two 300 s images each in V and B were obtained. Of these, however, the first had reduced count rates for all stars and a distorted point-spread function, possibly because the telescope and instrument had not yet settled when the exposure was started; we have not used that image. For calibration, sets of 30 s B and V images of the photometric standard field PG 1633+099 were acquired both before and after the science frames.

Following standard prescriptions, individual frames were bias-corrected and flat-fielded using twilight flats. Hot pixels and cosmic rays were replaced by a median over their neighbours. The instrumental fluxes were measured inside $3.''6$ radii and then corrected to a radius of $7''$ using measurements of bright isolated stars. For the calibration, we used 5 standard stars with a range of $B - V$ colors in the PG 1633+099 field (Stetson 1990). Measured magnitudes were compared to their catalogued counterparts to derive zero-points and colour terms. Both calibration sets yielded similar results. Small differences in airmass were corrected using standard values for La Silla. The root-mean-square residuals of the zero points in both bands were $\sim$0.01 mag. We find that the optical counterpart of the WD has $V = 21.30(5)$

and $B = [21.70(7), 21.73(7)]$, where the two measurements in B are for the two exposures, and where for the errors, we combined in quadrature the measurement and zero-point uncertainties. For the comparison star, we measure $V = 18.00(1)$ and $B = [18.73(2), 18.75(3)]$. Since the B magnitudes are consistent, we use the averages below.

We verified our calibration in several ways. First, we integrated our flux-calibrated spectra over the B-band filter curve of Bessell (1990). For the comparison star, using the wide-slit spectra, we find $B' = 18.81$ for the Gemini spectrum and $B = 18.71$ for the Keck spectrum. For the white dwarf, we find $B' = 21.69$ for the averaged Gemini white dwarf spectra, and $B = 21.70$ for the single narrow-slit Keck spectra. Here, we label the Gemini magnitudes as B', since the GMOS spectra do not fully cover the Bessell-B bandpass, which will introduce color terms.

Second, we tried to calibrate the g'-band GMOS acquisition images, by calibrating relative to our velocity standard, WD 1743−132, which has $V = 14.290$, $B - V = 0.300$ (Mermilliod et al. 1990), and thus, using the relations of Fukugita et al. (1996), $g' = V + 0.56(B - V) - 0.12 = 14.34$. We find $g' = 18.23$ for the comparison star and, using the average magnitude difference $\Delta g' = 3.091(17)$ between the WD and the comparison, we infer $g' = 21.32$ for the WD (here, the uncertainty will be dominated by systematics, but should be ≤ 0.05 mag). These numbers are consistent with the $g' = 18.39(7)$ and 21.42(7) expected from our SOAR photometry.

Looking at individual acquisition frames, the scatter of the magnitude difference was ~0.05 mag, somewhat larger than expected based on measurement noise, though with no obvious correlation with orbital phase. We find somewhat smaller scatter from convolving individual flux calibrated WD and comparison spectra with the Bessell B-band, and using those to determine differences (see Table 4.1). Ignoring the two points from our worst night (2006 May 26), the root-mean-square scatter is 0.032 mag. Since no obvious phase dependence is found, this places a limit on the irradiation of the WD atmosphere from the pulsar. However, the limit is too weak to be useful: Assuming a spin-down luminosity of $L_{\rm PSR} = dE/dt = -4\pi^2 I \dot{P}/P^3 \sim 4.8 \times 10^{33}\,{\rm ergs\,s^{-1}}$ and defining an irradiation temperature $T_{\rm irr} = (L_{\rm PSR}/4\pi a^2 \sigma)^{1/4} \simeq 3{,}800\,{\rm K}$ (where from Table 4.2, we inferred $a \simeq 1.8 \times 10^{11}$ cm), the expected orbital modulation is only $\Delta L/L \simeq [\pi R_{\rm WD}^2 (L_{\rm PSR}/4\pi a^2)/L_{\rm WD}] \sin i \simeq [T_{\rm irr}^4/4T_{\rm WD}^4] \sin i \leq 4 \times 10^{-3}$.

4.3 Results

4.3.1 Radial Velocities

Radial velocities of the WD, the comparison and the velocity standard were extracted by fitting their spectra with templates using the method discussed in Bassa et al. (2006). For the comparison, we first classified it using the on-line atlas by R. O. Gray.[1]

[1] http://nedwww.ipac.caltech.edu/level5/Gray/Graycontent.

We find that its spectrum resembles that of a G0V star, with an uncertainty of about 1 subtype. Comparing with various spectra from the UVESPOP[2] library of high resolution spectra (Bagnulo et al. 2003), we find the best fit for the G1V star HD 20807 (where, to match the resolution of the observations, we convolve the UVESPOP spectra with a Gaussian with FWHM equal to that of the seeing, truncated at the slit width). We fitted this template to each spectrum for a range of velocities, from -600 to $600\,\mathrm{km\,s^{-1}}$ with a step size of $5\,\mathrm{km\,s^{-1}}$. We corrected for the $11.5\,\mathrm{km\,s^{-1}}$ barycentric velocity of HD 20807 after the fact.

Similarly, the WD spectra were compared to an appropriate DA model atmosphere. The latter was determined iteratively, where we first fitted a high S/N single spectrum with a grid of model atmospheres created by one of us (D. Koester, see next section), then used the best fit solution to shift the spectra and average them at zero velocity, and finally fitted the average again to determine the best template. For WD 1743−132 we fitted the single spectrum with the grid and determined all parameters simultaneously.

For all above fits, we multiplied the templates with a 3rd degree polynomial to account for the normalization and possible variations with wavelength (see §4.5 for details). Our best fits gave typical reduced χ^2 values of $\chi^2_{\mathrm{red,min}} \sim 1.2$, 2.2 and 1.6 for the WD, the comparison star, and the velocity standard, respectively. Best-fit velocities were determined by fitting a parabola to the χ^2 values to within $60\,\mathrm{km\,s^{-1}}$ of minimum, with uncertainties taken to be the difference in velocity over which χ^2 increased by $\chi^2_{\mathrm{red,min}}$ (thus effectively increasing our uncertainties to account for the fact that $\chi^2_{\mathrm{red,min}}$ did not equal unity).

For the Keck spectra, we proceeded similarly. Here on the red side, we could not use the UVES spectrum due to a gap over the Ca II triplet, and hence we used instead a $T_{\mathrm{eff}} = 6{,}000\,\mathrm{K}$, $\log g = 4.5\,\mathrm{dex}$ model by Zwitter et al. (2004). As we trust the absolute wavelength calibration of this observation most (being calibrated relative to telluric emission lines), we use this estimate of the velocity below to transform all velocities to the barycentric reference frame.

4.3.2 Radial Velocity, Orbit and Mass Ratio

In Table 4.1 we list the measured radial velocities for all targets, with barycentric corrections applied. For determining the orbit, we folded the barycentric velocities using the ephemeris in Table 4.2 and fitted for a circular orbit keeping the orbital period and time of accenting node passage fixed to the timing values. The fit gave a radial velocity semi-amplitude of $K_{\mathrm{obs}} = 165 \pm 7\,\mathrm{km\,s^{-1}}$ and a systemic radial velocity of $\gamma = -50 \pm 4\,\mathrm{km\,s^{-1}}$ with $\chi^2_{\mathrm{red}} = 1.55$ for 16° of freedom.

The radial velocity of the comparison star in the Gemini dataset varied as much as $55\,\mathrm{km\,s^{-1}}$ which is considerably higher than the uncertainties of individual points. We found no evidence for binarity and thus we attribute the large scatter to systematics,

[2] http://www.sc.eso.org/santiago/uvespop/DATA.

Table 4.2 Parameters for the PSR J1738+0333 system

Timing parameters	
Reference time (MJD)	54600.0001776275
Right ascension, α (J2000)	$17^{h}\ 38^{m}\ 53^{s}9658386$ (7)
Declination, δ (J2000)	$03.^{\circ}\ 33'\ 10.''86667$ (3)
Proper motion in α, μ_α (mas yr^{-1})	+7.037 (5)
Proper motion in δ, μ_δ (mas yr^{-1})	+5.073 (12)
Parallax, π_x (mas) (a)	0.68 (5)
Spin frequency, ν (Hz)	170.93736991146392(3)
First derivative of ν, $\dot{\nu}$ (fHz s^{-1})	−0.704774 (4)
Orbital period P_b (days)	0.3547907398724(13)
Projected semi-major axis, x (lt-s)	0.343429130 (17)
Time of ascending node, $T_{\rm asc}$ (MJD)	54600.20040012(5)
$\eta \equiv e \sin \omega$	$(-1.4 \pm 1.1) \times 10^{-7}$
$\kappa \equiv e \cos \omega$	$(3.1 \pm 1.1) \times 10^{-7}$
First derivative of P_b, $\dot{P}_b$ (fs s^{-1})	−17.0 (3.1)
"range" parameter of Shapiro delay, r (μs) (b)	0.8915
"shape" parameter of Sapiro delay, $s \equiv \sin i$ (b)	0.53877
Dispersion measure, DM (cm^{-3} pc)	33.77312 (4)
Test parameters	
First derivative of x, $\dot{x}$ (fs s^{-1})	0.7(5)
Second derivative of ν, $\ddot{\nu}$ (10^{-28} Hz s^{-2})	−0.6 (2.3)
Derived parameters	
Galactic Longitude, l	27°.7213
Galactic Latitude, b	17°.7422
Distance, d (kpc)	1.47 (10)
Total proper motion, μ (mas yr^{-1})	8.675 (8)
Position angle of proper motion, Θ_μ (J2000)	53°.72 (7)
Position angle of proper motion, Θ_μ (Galactic)	116°.12(7)
Spin period, P (s)	0.0058 50095859775683 (5)
First derivative of spin period, $\dot{P}$ (10^{-20}s s^{-1})	2.411991 (14)
Intrinsic $\dot{P}$, $\dot{P}_{\rm Int}$ (10^{-20}s s^{-1}) (a)	2.243 (13)
Characteristic age, τ_c (Gyr)	4.1
Transverse magnetic field at the poles, B_0 (10^9G)	0.37
Rate or rotational energy loss, $\dot{E}$ (10^{33} erg s^{-1})	4.4
Mass function, f ($M_\odot$)	0.0003455012 (11)
Mass ratio, $q \equiv M_p/M_c$	8.1 (2)
Orbital inclination, i (°)	32.6 (1.0)
Temperature (K)	9130 (150)
Surface gravity ($\log g$, spectroscopy)	6.55 (10)
Surface gravity ($\log g$, $\dot{P}_b + q + \pi$+photometry)	6.45 (7)
Photometry, V-band	21.30 (5)

(continued)

Table 4.2 continued

Photometry, B-band	21.71 (4)
Semi-amplitude of radial velocity, $K_{\rm WD}$ (km s^{-1})	171 (5)
Systemic radial velocity, γ (km s^{-1})	−42 (16)
Transverse velocity, v_T (km s^{-1})	59 (6)
3D velocity amplitude (km s^{-1})	72 (17)
Mass ratio, q	8.1 (2)
WD mass, $M_{\rm WD}$ (M$_\odot$, spectroscopy)	$0.181^{+0.007}_{-0.005}$
WD mass, $M_{\rm WD}$ (M$_\odot$, q + $\dot{P}_{\rm b}$)	0.182 ± 0.016
WD radius (Spectroscopy) (R$_\odot$)	$0.037^{+0.004}_{-0.003}$
WD radius (Photometry) (R$_\odot$)	0.042 (4)
Cooling age, $\tau_{\rm c}$ (Gyr)	0.5 – 5
Pulsar mass, M_p (M$_\odot$)	$1.46^{+0.06}_{-0.05}$
Total mass of binary, M_t (M$_\odot$)	$1.65^{+0.07}_{-0.06}$
Eccentricity, e	$(3.4 \pm 1.1) \times 10^{-7}$
Apparent $\dot{P}_b$ due to Shklovskii effect, $\dot{P}_b^{\rm Shk}$ (fs s^{-1}) (a)	$8.2^{+0.6}_{-0.5}$
Apparent $\dot{P}_b$ due to Galactic acceleration, $\dot{P}_b^{\rm Gal}$ (fs s^{-1}) (a)	$0.58^{+0.16}_{-0.14}$
Intrinsic $\dot{P}_b$, $\dot{P}_b^{\rm Int}$ (fs s^{-1}) (a)	−25.9 (3.2)
Predicted $\dot{P}_b$, $\dot{P}_b^{\rm GR}$ (fs s^{-1})	$-27.7^{+1.5}_{-1.9}$
"Excess" orbital decay, $\dot{P}_b^{\rm xs} = \dot{P}_b^{\rm Int} - \dot{P}_b^{\rm GR}$ (fs s^{-1}) (a)	$+2.0^{+3.7}_{-3.6}$
Time until coalescence, τ_m (Gyr)	∼13.2

In parentheses we present the 1-σ uncertainties in the last digit quoted, as estimated by TEMPO2. If the value and uncertainty are signalled with an (a) then they were derived from a Monte-Carlo procedure (Freire et al. 2012). (b) The Shapiro delay parameters r and s were not fitted in the derivation of the timing model; the values used were derived from a combination of other timing and optical parameters (Freire et al. 2012). All timing parameters are derived using TEMPO2 and are displayed as measured at the Solar System Barycenter, in barycentric coordinate time (TCB). The "test parameters" were not fitted when deriving the main timing model, but their values were derived fitting for all the other parameters in the model

likely induced by slit positioning errors and differential atmospheric diffraction. For that reason, we chose to use velocities relative to the comparison star, Δv. This choice relies on the assumption that both the WD and the comparison star are affected by the same systematics. This should be correct to first order, but given the relatively large separation of the two stars on the slit, their different distances from the centre of rotation of the instrument, and their different colours, small second-order differences may remain. Even if any are present however, they should not be correlated with orbital phase (since our measurements are taken on many different nights), and thus be taken into account automatically by our rescaling of the measurement errors such that reduced χ^2 equals unity.

After subtracting the velocity of the comparison star, we obtain $K_{\rm obs} = 166 \pm 6\,{\rm km\,s^{-1}}$, $\Delta\gamma = -101 \pm 4\,{\rm km\,s^{-1}}$ with $\chi^2_{\rm red} = 1.07$. This orbit is shown in Fig. 4.1. This fit has two outliers, which both are from spectra taken in the night with the

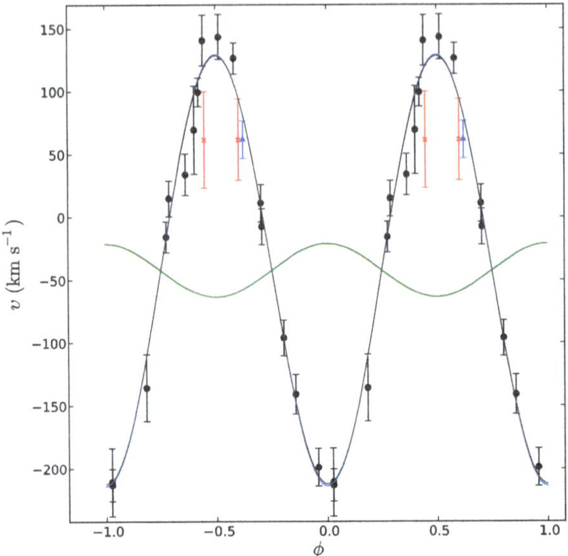

Fig. 4.2 Radial velocity measurements of the companion to PSR J1738+0333 as a function of the orbital phase. *Filled black circles* depict the points used to fit the orbit and the *blue line* the best-fit solution. *Red crosses* indicate two outliers that we excluded and the *black line* the best-fit solution with these points included. The latter agrees well and is almost indistinguishable. The *blue triangle* shows the Keck point. The *green line* depicts the velocity of the pulsar as inferred from radio timing. All velocities are relative to the comparison star, but corrected for its estimated 61 km s^{-1} barycentric radial velocity. All *error bars* represent 1σ uncertainties. The orbit is depicted two times for clarity

worst condition (they are also outliers in the relative flux between the WD and the comparison star; see Table 4.1). Excluding these, we find $K_{\rm obs} = 167 \pm 5\,{\rm km\,s^{-1}}$ and $\Delta\gamma = -103 \pm 3\,{\rm km\,s^{-1}}$ with $\chi^2_{\rm red} = 0.93$ for 14° of freedom. We will use these latter values as our best estimates, but note that all fits gave consistent results, so our inferences do not depend on this choice.

Because the exposure time is a significant fraction of the orbit ($t_{\rm exp} \simeq 0.12\,P_{\rm b}$), the observed semi-amplitude is affected by velocity smearing. This reduces the measured amplitude by a factor $\sin(\pi t_{\rm exp}/P_{\rm b})/(\pi t_{\rm exp}/P_{\rm b}) = 0.976$. Thus, the true radial-velocity amplitude is $K_{\rm WD} = 171 \pm 5\,{\rm km\,s^{-1}}$.

Likewise, the semi-amplitude of the pulsar's projected radial velocity is $K_{\rm PSR} = 2\pi c x/P_{\rm b} = 21.103059(2)\,{\rm km\,s^{-1}}$, where x is the projected semi-major axis of the pulsar orbit. Based on the two values calculated above we derive a mass ratio of $q = K_{\rm WD}/K_{\rm PSR} = 8.1 \pm 0.2$.

4.3.3 Systemic Velocity

The systemic velocity $\Delta\gamma$ derived above is relative to the comparison star. Thus, for an absolute value one needs to obtain an estimate of the true velocity of the latter. From the Gemini spectra we derived an average value of 64 ± 5 km s^{-1}. As discussed above, the individual velocities have a large scatter and one may thus worry about systematics. It seems, that these are of the order of $15 - 20$ km s^{-1}. First, for the velocity standard WD 1743−132 we find a velocity of -58.6 ± 1 km s^{-1}, which is offset by 14.2 km s^{-1} from the catalogue value of -72.8 km s^{-1} Reid (1996). Second, for the comparison star, our Keck spectrum yields 61 ± 5 km s^{-1}. As mentioned above, we believe the wavelength calibration is most reliable for the Keck spectrum, so we adapt this velocity. For PSR J1738+0333, correcting for the gravitational red-shift of the white dwarf of 3 km s^{-1} (using the mass and radius derived in Sect. 4.3), we infer a systemic velocity of $\gamma = -42 \pm 16$ km s^{-1}.

4.3.4 Interstellar Reddening

We calculated the run of reddening along the line of sight using the Galactic extinction model of Drimmel et al. (2003a). We find that the interstellar extinction increases smoothly to reach a maximum value of $A_V = 0.56$ at 1.3 kpc and remains constant thereafter. This is similar to the maximum value along this line of sight of $A_V = 0.65$ inferred from the maps of Schlegel et al. (1998a). Therefore, for both PSR J1738+0333 and the comparison value we adopt $A_V = 0.56 \pm 0.09$, with the uncertainty taken to be the difference between the two models.

We can now use these results to estimate the distance of the comparison star: Adopting $M_V = 4.3$ and $(B - V)_0 = 0.57$ for a G0V star Cox (2000) and $A_B = 1.321 A_V$ Schlegel et al. (1998a) we obtain a distance of ~4.3 kpc for both bands. As a sanity check for the systemic velocity derived above, we can calculate the expected velocity of the comparison for the photometric parallax: Assuming the Galactic potential of Kenyon et al. (2008), a distance to the Galactic center of 8.0 kpc and a peculiar velocity of the Sun relative to the local standard of rest of $(U, V, W) = (10.00, 5.25, 7.17)$ km s^{-1} Cox (2000), we find that the local standard of rest at the position of the comparison star moves with a speed of ~60 km s^{-1}. Given the uncertainties of the model and our measurements and the possibility of peculiar motion, the latter agrees well with our estimated value.

4.3.5 Temperature and Surface Gravity of the White Dwarf

The zero-velocity average spectrum (Fig. 4.3) shows deep Balmer lines up to H12, typical for a WD with a hydrogen atmosphere and low surface gravity.

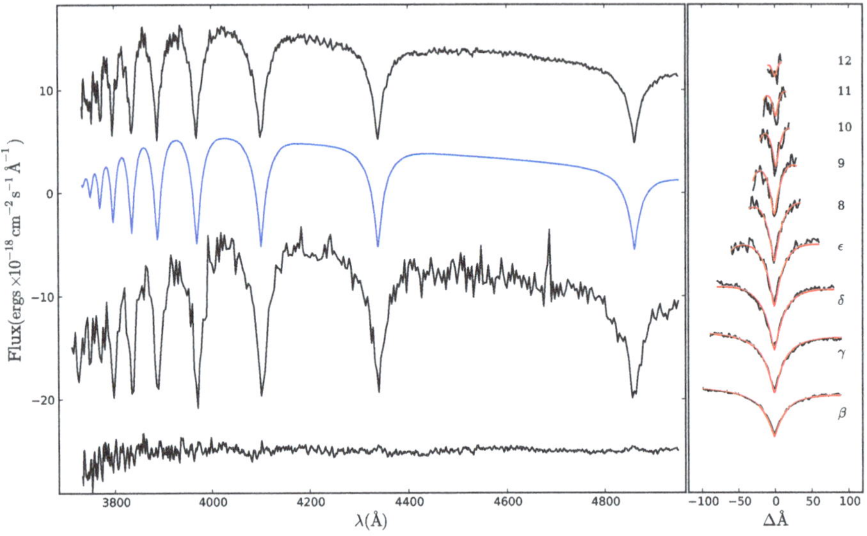

Fig. 4.3 *Left* From *top* to *bottom*: The zero-velocity flux-calibrated average spectrum of PSR J1738+0333 obtained with Gemini, the corresponding best-fit atmospheric model, the (single) spectrum obtained with Keck, and the residuals from the fit (see Sect. 4.3). The model and the Keck spectrum are shifted down by 10 and 20 units respectively. *Right* Details of of the Balmer series in the average spectrum (Hβ to H12, from *bottom* to *top*), with the best-fit model overdrawn (*red lines*). *Lines* are shifted by 8 units with respect to each other

Quantitative estimates for the atmospheric parameters were obtained by modelling the spectrum with a grid of DA model atmospheres extending from 7,000 K to 25,000 K and $\log g = 6.00$ to $\log g = 8.00$ with step-sizes of 100 K and 0.1 dex respectively. The models used in this chapter are a recent update of the grid presented in Koester (2008) which incorporates the improved treatment of pressure broadening of the absorption lines by Tremblay and Bergeron (2009).

At each point of the grid that we scanned, we fitted for the normalization with a polynomial function of the wavelength. This was found necessary in order to account for the (up to) ~10 %, slowly varying continuum deviations, caused by in-perfect flux calibration. Assuming our flux calibration is perfect (namely, using a normalization factor that does not vary with wavelength) resulted in a poor fit with large scale structure in the residuals and lines systematically deeper than the best-fit model (best-fit values: $T_{\text{eff}} = 9010 \pm 50\,\text{K}$, $\log g = 6.81 \pm 0.12\,\text{dex}$ with $\chi^2_{\text{red}} \sim 9$). Similarly underestimated lines were obtained using a fitting routine normally used by one of us (D. Koester) that assumes a fixed slope for the continuum over the length of each line. The former comparison revealed that there was also a smaller spectral range between 4,400 – 4,780 Å with features similar with the ones seen in the flat fields (see Sect. 4.2.2), likely associated with the holographic grating (we were alerted to this effect because it was much more obvious for the companion of PSR J1909−3744; vK+12). Fortunately, no Balmer lines are present in this region,

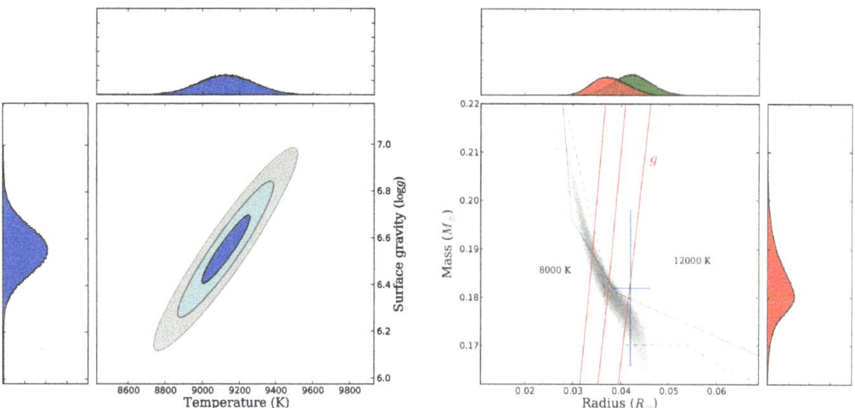

Fig. 4.4 *Left* Constraints on the temperature and gravity of the white dwarf companion to PSR J1738+0333 inferred from our model-atmosphere fit, with contours at $\Delta\chi^2 = \chi^2_{\rm red,min}$, $4\chi^2_{\rm red,min}$, $9\chi^2_{\rm red,min}$, and $16\chi^2_{\rm red,min}$. The horizontal and *vertical sub-panels* show the histograms of the distributions for $T_{\rm eff}$ and $\log g$ from our Monte Carlo simulation. *Right* Constraints on the mass and radius of the WD. The *shaded area* depicts the distribution of realizations from our Monte-Carlo simulation. Overdrawn are: the central value and 1σ confidence limits of the observed surface gravity (*red lines*); the model tracks of Panei et al. (2000) for constant temperature (8,000 and 12,000 K; *dotted lines*); the mass-radius relations of Serenelli et al. (2001) (*solid*) and Panei et al. (2007) (*dashed*) for our best-fit temperature of 9,130 K; *errorbars* showing the independent constraints from photometry and radio timing. The horizontal and vertical panel show the inferred distributions for the WD radius and mass, respectively, as well as the independent photometric estimate for the former (in *green*; see also text in Sect. 4.3.5)

and hence we simply modelled the spectrum excluding this range (specifically, we fitted the ranges 3,700–4,400, and 4,780–4,960 Å). Like for our radial-velocity fits, we accounted for the spectral resolution by convolving the models with a truncated Gaussian.

Using the choices described above we obtain $T_{\rm eff} = 9129 \pm 20$ K (implying a spectral type DA5.5) and $\log g = 6.55 \pm 0.07$ dex with $\chi^2_{\rm red,min} \simeq 1.5$ (for ~800 points and 6 parameters). Here, the best-fit values and statistical uncertainties were determined by fitting the χ^2 surface with a paraboloid as in Bassa et al. (2006). We verified these estimates using a Monte-Carlo simulation with 10^6 iterations (see Fig. 4.4). The results are almost identical, with the simulation giving slightly larger uncertainties. However, as we will see below, the systematic uncertainties are larger.

The best-fit model is shown in Fig. 4.3. Most lines are matched almost perfectly, but H11 and H12 are slightly underestimated. We do not know the reason for this. As the continuum matches very well, it cannot be due to errors in the flux calibration (which would be multiplicative), while most other observational issues (scattered light, etc.) would lead to lines that have reduced rather than increased depth.

Given the above discrepancies, as well as previous experience with fitting model atmospheres, it is likely our uncertainties are dominated by systematics rather than measurement errors. We investigated this in three ways. First, we tried small changes in the assumed spectral resolution (by 5 %) and varied the different polynomial degrees for the continuum (2nd to 4th order). The former had only very small effect ($\sim$20 K changes in T and $\sim$0.03 dex changes in $\log g$), while changing the degree of the polynomial caused larger differences: 0.1 dex (1.5σ) for the surface gravity and up to 150 K (7σ) for the temperature. Our central values are based on a 3rd degree polynomial, since it gave the best fit for the higher lines.

As a second check, we obtained an independent measure of the atmospheric parameters using the Keck spectrum. Again using a third-degree polynomial for the continuum, and fitting the same wavelength regions, we find $T_{\rm eff} = 9{,}281 \pm 110$ K and $\log g = 6.57 \pm 0.13$ dex. Here, switching between polynomials for the continuum normalization had a slightly smaller impact on the estimated values ($\sim$100 K in T and $\sim$0.1 in $\log g$). While the surface gravity agrees almost perfectly with the Gemini value, the effective temperature is somewhat higher, suggesting, again, that temperature is more sensitive to our modelling assumptions.

Finally, we fitted the individual spectra with the model atmospheres and obtained a mean temperature of $\langle T \rangle = 9{,}153 \pm 38$ K with an rms scatter of 155 K.

From the above, it is clear the formal uncertainty on especially the temperature is too small, and we adopt as realistic estimates $T_{\rm eff} = 9{,}130 \pm 150$ K and $\log g = 6.55 \pm 0.10$ dex. Fortunately, the effect of the larger temperature uncertainty on the derived masses is small, because the mass-radius relation is much more sensitive to surface gravity than to temperature. For our mass calculation below, we thus choose to inflate the original χ^2 map to include the systematics mentioned above but preserve information about the covariance between parameters.

Finally we searched the average spectrum for signatures of rotational broadening. For that we proceeded in two ways: First, we broadened a 9,000 K, $\log g = 6.5$ model atmosphere using the analytical profile of Gray (2005) with a limb darkening coefficient of 0.3 and scanned a grid of rotational velocities $0 \leq v_{\rm r} \sin i \leq 1{,}500\,{\rm km\,s^{-1}}$ in steps of $50\,{\rm km\,s^{-1}}$. Second, we let all parameters free. In both cases we accounted for the spectral resolution of the instrument as above. We find the rotational broadening consistent with zero with the 1σ upper limit being 440 and $510\,{\rm km\,s^{-1}}$ respectively.

4.3.6 White Dwarf Radius from Photometry

We can use the best-fit atmosphere model, the observed fluxes, and the distance to obtain an estimate of the WD radius. In terms of magnitudes,

$$m_\lambda - 5\log(d/10\,{\rm pc}) - A_\lambda = 43.234 - 5\log(R/{\rm R}_\odot) - 2.5\log {\rm F}_\lambda + {\rm c}_\lambda \qquad (4.1)$$

where m_λ is the apparent magnitude in band λ, the numerical term is $-5\log({\rm R}_\odot/10\,{\rm pc})$, F_λ is the emitted flux per unit surface area integrated over the relevant filter, and c_λ the zero-point. Convolving the best-fit model with the B and V band passes of Bessell (1990) yields $F_B = 6.289 \times 10^7\,{\rm erg\,cm^{-2}\,s^{-1}\,\AA^{-1}}$ and

$F_V = 4.353 \times 10^7\,\mathrm{erg\,cm^{-2}\,s^{-1}\,\AA^{-1}}$. Here the uncertainty due to the fit is ~5 % (mostly due to the ~1.5 % uncertainty in temperature). Using the zero-points of Bessell (1990), $c_B = -20.498$ and $c_V = -21.100$, and the reddening inferred above we obtain radii $R = 0.042 \pm 0.004\,\mathrm{R}_\odot$ and $R = 0.042 \pm 0.004\,\mathrm{R}_\odot$ for B and V, respectively (with the uncertainty dominated by the uncertainty in the parallax).

4.3.7 Masses of the White Dwarf and the Pulsar

The mass of the WD can be estimated using a mass-radius relation appropriate for low mass helium white dwarfs. We use the finite-temperature relation for low-mass WDs from Panei et al. (2000), which gave good agreement for the companion of PSR J1909−3744 (vK+12).

For the calculation we proceeded as follows: We sampled the inflated χ^2 surface derived above in a Monte-Carlo simulation using 10^6 points uniformly distributed in the $T_{\mathrm{eff}} - \log g$ plane. For each point within the expectations, we linearly interpolated the 8000 and 12000 K models of Panei et al. (2000) for WDs with extended hydrogen envelopes to the given temperature and calculated the mass and radius at the cross-section of the observed value (which scales as $g = GM/R^2$) and the model. Subsequently, we calculated the mass of the pulsar, assuming a normal distribution for the mass ratio with $q = 8.1 \pm 0.2$. Furthermore, we calculated the inclination using the mass function f_{M} of the binary ($\sin^3 i = f_M (M_{\mathrm{WD}} + M_{\mathrm{PSR}})^2 / M_{\mathrm{WD}}^3$).

We show the mass distribution in Fig. 6.4. Since the mass-radius relation is steeper towards higher masses, the companion's mass distribution is asymmetric, with larger wings towards higher masses. The same holds for the distribution for the radius, with larger wings towards smaller radii. The error on the pulsar mass is dominated by the uncertainties in the companion's mass estimate. To summarize, the values that we will be using for the rest of this chapter are: $M_{\mathrm{WD}} = 0.181^{+0.007,+0.017}_{-0.005,-0.013}\,\mathrm{M}_\odot$, $M_{\mathrm{PSR}} = 1.47^{+0.07,+0.14}_{-0.06,-0.08}\,\mathrm{M}_\odot$, $R_{\mathrm{WD}} = 0.037^{+0.004,+0.007}_{-0.003,-0.006}\,\mathrm{R}_\odot$ and $i = 32.6^{\mathrm{o}+1.0,+2.1}_{-1.0,-2.1}$. Here, the errors separated by commas are the corresponding 68 and 95 % intervals spanned by the Monte-Carlo realizations.

Finally, we also derived mass estimates using two different sets of tracks, that gave reliable results for PSR J1909−3744 (vK+12): The tracks of Serenelli et al. (2001) yielded $M_{\mathrm{WD}} = 0.183^{+0.007,+0.011}_{-0.004,-0.005}\,\mathrm{M}_\odot$ and $R_{\mathrm{WD}} = 0.037^{+0.005,+0.007}_{-0.004,-0.007}\,\mathrm{R}_\odot$, almost identical to the above. The tracks of Panei et al. (2007) yielded slightly different values: $M_{\mathrm{WD}} = 0.175^{+0.017+,0.029}_{-0.005,-0.006}\,\mathrm{M}_\odot$ and $R_{\mathrm{WD}} = 0.038^{+0.005,+0.010}_{-0.003,-0.004}\,\mathrm{R}_\odot$. However, we note that these models predict a cooling age much smaller than the characteristic age of the pulsar (see next section).

4.3.8 Cooling Age

We compared the absolute photometric magnitudes in B and V with the theoretical cooling tracks of Serenelli et al. (2001) for solar metallicity progenitors to infer the cooling age of the WD. We did this by minimizing a χ^2 merit function based on the sum of differences between observed and model fluxes in both bands. The track of Serenelli et al. (2001) closest in mass to the companion of PSR J1738+0333 is that of a $0.169\,M_{\odot}$, for which we find $\tau_c \sim 4.2$ Gyr. For that age and mass, the predicted temperature and surface gravity are $T_{\rm eff} \sim 8{,}500$ K and $\log g \sim 6.35$ dex. For our best-fit spectroscopic estimates the same track yields $\tau_c \sim 2.6$ Gyr. Since the observed mass is slightly heavier, its cooling age must be somewhat lower. Using the $0.193\,M_{\odot}$ track, we get $\tau_c \sim 600$ Myrs. The large difference is due to the dichotomy around $0.2\,M_{\odot}$ expected between WDs with thick and thin hydrogen atmospheres. Using the tracks of Panei et al. (2007), for the mass of $0.175\,M_{\odot}$ inferred using those, we again find short ages, $\tau_c \sim 500$ Myrs from the photometry and $\tau_c \sim 450$ Myrs for the spectroscopic parameters.

Finally, the suggested relatively large age of the system (4 Gyr plus 2–10 Gyr for the progenitor to have evolved) motivated us to compare our observations with models for lower metallicity progenitors. Using the $0.183\,M_{\odot}$, $Z = 0.001$ track of Serenelli et al. (2002) we obtain $\tau_c \sim 5$ Gyr.

The above analysis demonstrates that with the current set of observations it is difficult to constrain the cooling age of the WD, since this depends on both the thickness of the WD envelope and the metallicity of its progenitor. Future, more precise constraints on the parallax and consequently on the radius, might help to discriminate between different cases.

4.3.9 3D Velocity and Galactic Motion

In Sect. 4.2 we computed the two components of the transverse velocity based on the parallax and proper motion estimates from radio timing measurements of the pulsar. Combined with the systemic radial velocity $\gamma = -42 \pm 16\,\mathrm{km\,s^{-1}}$ from the optical observations of the white dwarf (Table 4.1), we have the full 3D velocity and can compute the Galactic path back in time [like was done for PSR J1012+5307 by Lazaridis et al. (2009a)]. For our calculations we have used the Galactic potential of Kenyon et al. (2008), verifying our results with those of Kuijken and Gilmore (1989) and Paczynski (1990). We infer that the PSR J1738+0333 system has an eccentric orbit with a Galacto-centric distance between 6 and 11 kpc, and an oscillating Z-motion with an amplitude of 1 kpc and a (averaged) period of 125 Myr. We also calculated the peculiar velocity of the system with respect to the local standard of rest at every transition of the Galactic plane ($Z = 0$) during the last 4 Gyr, and find that it ranges between 70 and $160\,\mathrm{km\,s^{-1}}$. We will discuss this further in Sect. 4.4.

4.4 Ramifications

In Table 4.2 we list the properties of the system derived in previous sections and in Fig. 4.5 we show our constrains on the masses. In what follows we discuss the ramifications of our work for stellar and binary astrophysics.

4.4.1 Kinematics

PSR J1738+0333 has a velocity of $85 \pm 17\,\mathrm{km\,s^{-1}}$ with respect to the local standard of rest that co-rotates with the Galaxy ($Z = 0$) at the distance of the pulsar. The latter compares well with the mean transverse velocity for the bulk of MSPs with measured proper motions [$\sim 85\,\mathrm{km\,s^{-1}}$ according to Hobbs et al. (2005)]. Our semi-quantitative analysis in Sect. 4.3 shows that the system's velocity varies as much as $150\,\mathrm{km\,s^{-1}}$ over the course of its Galactic orbit. Based on the simplified potential of Kenyon et al. (2008) used herein, PSR J1738+0333 has a peculiar velocity between 70 and $160\,\mathrm{km\,s^{-1}}$ when it crosses the Galactic plane ($Z = 0$). Thus, assuming that the system had a small peculiar motion before the SN explosion, the systemic velocity after the formation of the NS must have been in that range. This is consistent

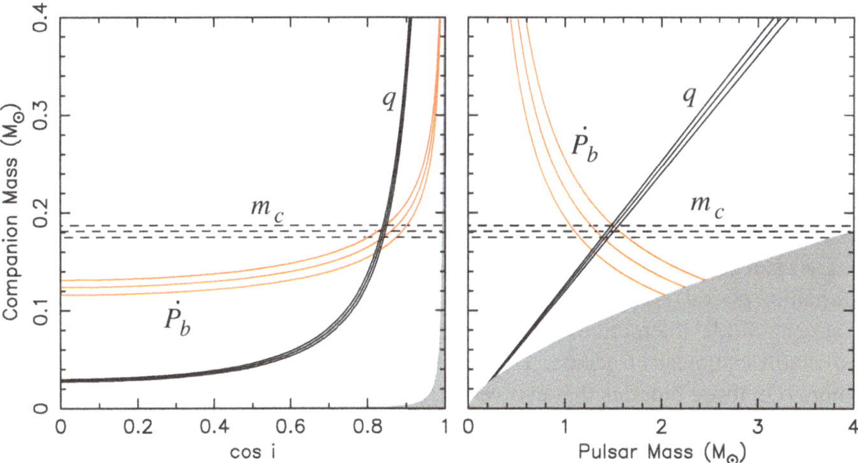

Fig. 4.5 Constraints on system masses and orbital inclination from radio and optical measurements of PSR J1738+0333 and its WD companion. The mass ratio q and the companion mass m_c are theory-independent (indicated in *black*), but the constraints from the measured intrinsic orbital decay ($\dot{P}_b^{\mathrm{Int}}$, in *orange*) are calculated *assuming* that GR is the correct theory of gravity. All curves intersect, meaning that GR passes this important test. *Left* $\cos i$–m_c plane. The *gray* region is excluded by the condition $m_p > 0$. *Right* m_p–m_c plane. The *gray* region is excluded by the condition $\sin i \leq 1$. Each triplet of curves corresponds to the most likely value and standard deviations of the respective parameters

with a SN explosion with a small, or even negligible kick (Tauris and Bailes 1996; Nice and Taylor 1995).

4.4.2 Evolutionary History

Millisecond pulsars with low-mass helium WD companions are expected to form through mainly two different channels depending on the initial separation of the progenitor binary (e.g. Tauris (2011) and references therein). The initial separation of the progenitor binary determines the evolutionary status of the donor star at the onset of the Roche lobe overflow (RLO):

- *Case A RLO*: For systems with initial periods short enough to initiate mass transfer on the main-sequence, it is expected that magnetic-braking (aided to some extent by gravitational radiation) drives the system to shorter periods, resulting in a compact binary in an orbit which is close to being perfectly circular (the eccentricity, $e < 10^{-5}$). These systems were first studied in detail by Pylyser and Savonije (1989).
- *Case B RLO*: For progenitors with larger initial separations the mass transfer is expected to start at a later phase, since the star fills its Roche lobe only during shell hydrogen burning, while moving-up the red giant branch. In this case the orbit will diverge resulting in a wider binary. Interestingly, for systems following this path, there are two theoretical predictions that can be verified observationally: The first is a correlation between the orbital period and the mass of the WD companion which results from the unique relation between the radius of the giant donor and the mass of its core which eventually forms the WD Savonije (1987). The second is a correlation between the orbital eccentricity and the orbital period Phinney (1992) arising because the turbulent density fluctuations in the convective envelope—of which the size increases in more evolved stars (wider orbits)—do not allow for a perfect tidal circularization.

The critical period that separates diverging from converging systems (often called bifurcation period) is expected to be ∼1 day, however its precise value depends on the treatment of tidal interactions and magnetic braking (e.g. Pylyser and Savonije 1989) and is still a subject of debate. The residual eccentricity in binaries with radiative donors (i.e. those binaries that evolve to tight converging systems) should be closer to zero compared to binaries in wider orbits but it is difficult to estimate by how much, as pointed out by Phinney and Kulkarni (1994).

With a current orbital period of 8.5 h, PSR J1738+0333 is most likely the fossil of the former case (Case A RLO). However it is interesting to note that our mass estimate and the non-zero eccentricity derived in Paper II (see also below) pass both tests for the latter case mentioned above (Case B RLO) that predict $m_c = 0.18 \pm 0.01\,\mathrm{M}_\odot$ Tauris and Savonije (1999) and $e \sim 4 \times 10^{-7}$, respectively (deduced by extrapolating the Phinney (1992) relation to the observed period). This apparent agreement seems to be confirmed not only for PSR J1738+0333 but also for the other short-period LMWD binaries with measured masses [PSR J1012+5307, Lazaridis et al.

(2009a); PSR J0751+1807, Nice et al. (2008)], as well as low-mass WD companions to non-degenerate stars [e.g., (van Kerkwijk et al. 2010b; Breton et al. 2011)]. Since companion masses in converging systems are not expected to follow these relations, we cannot exclude a coincidence, but the matches seem to suggest that there is a grey zone with properties from both cases—something which should help improve our as yet rather simplified models of these systems.

4.4.3 Pulsar Mass and Efficiency of the Mass Transfer

Regardless of the evolutionary path followed, the mass transfer was sub-Eddington (e.g. Tauris and Savonije 1999) and thus one would expect that a substantial fraction of the mass leaving the donor was accreted by the neutron star. For PSR J1738+0333, this is demonstrably false: The minimum mass of the donor star can be constrained from our WD mass estimate to be $\geq 1\,\mathrm{M}_{\odot}$ because the available time for evolution is limited by the Hubble time (minus the cooling age of the WD). The amount of mass lost by the donor is $M_{\rm donor} - M_{\rm WD}$, while the amount accreted by the pulsar is $M_{\rm PSR} - M_{\rm PSR}^{\rm init}$, with the last term being the birth mass prior to accretion. For any realistic birth mass of the neutron star at the low end of its "canonical" birth mass range ($\geq 1.20\,M_{\odot}$), we find that more than 60 % of the in-falling matter must have escaped the system (after correcting for the conversion from baryonic mass to gravitational mass). This translates to an accretion efficiency of only $\varepsilon < 0.40$. This result confirms the findings of Tauris and Savonije (1999) who concluded that a substantial fraction of the transferred matter in LMXBs is lost from the system, even at sub-Eddington mass-transfer rates.

Possible mechanisms for mass ejection discussed in the literature include propeller effects, accretion disc instabilities and direct irradiation of the donor's atmosphere from the pulsar (e.g., Illarionov and Sunyaev 1975). Alternatively, the neutron star in PSR J1738+0333 might have formed via the accretion-induced collapse of a massive ONeMg WD. If the neutron star was formed towards the end of the mass transfer it would not have accreted much since its birth. A possible problem with the above mechanism however, is that it is specific to pulsars, while similarly inefficient accretion has been found also for low-mass WDs with non-degenerate companions (e.g., Breton et al. 2011), suggesting the problem in our understanding is more general. Finally, we note that even major inefficiencies in the mass accretion process do not pose a problem for the recycling scenario: the accreted mass needed to spin-up a pulsar to a $\sim$5 ms period is only of the order of $0.05\,\mathrm{M}_{\odot}$.

More constraining (but less stringent) estimates are also obtained for the 6.3 h orbital period binary, PSR J0751+1807 Nice et al. (2008) for which we find $\varepsilon \sim 0.1 - 0.3$.

4.5 Conclusions

The main result of this chapter is the determination of the component masses of the PSR J1738+0333 system and adds to the three previously known MSP-LMWD binaries with spectroscopic information [PSR J1012+5307, van Kerkwijk et al. (1996) and Callanan et al. (1998);PSR J1911−5958A, Bassa et al. (2006);PSR J1909 −3744, vK+12].

Our mass estimates are derived independently of any strong field effects and thus transform the PSR J1738+0333 system into a gravitational laboratory, which—due to its short orbital period, gravitationally asymmetric nature, and timing stability—provides the opportunity to test the radiative properties of a wide range of alternatives to GR (see Paper II for details).

Based on our measurements of the component masses, GR predicts an orbital decay of $\dot{P}_\mathrm{b} = -2.77^{+0.15}_{-0.19} \times 10^{-14}$. While the actual $\dot{P}_\mathrm{b}$ inferred observationally is still less precise than this prediction, it will eventually provide a precise test for the input physics of atmospheric and evolutionary models. Assuming the validity of GR, one can confront the spectroscopic WD mass estimate implied by the mass ratio and intrinsic orbital decay of the system and thus test the assumptions for stellar astrophysics and WD composition that were used to model the evolution of the WD. Additionally, this mass estimate, combined with parallax and absolute photometry constrains independently the surface gravity of the WD. The current estimates on these parameters imply a surface gravity of $\log g = 6.45 \pm 0.07$ dex. While this is

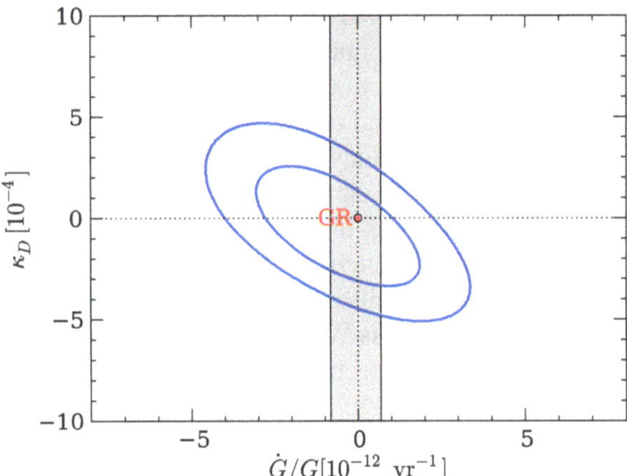

Fig. 4.6 Limits on $\dot{G}/G$ and κ_D derived from the measurements of $\dot{P}_b^{\mathrm{xs}}$ of PSR J1738+0333 and PSR J0437−4715. The *inner blue contour* level includes 68.3 % and the outer contour level 95.4 % of all probability. At the origin of coordinates, general relativity is well within the inner contour and close to the peak of probability density. The *gray band* includes regions consistent with the measured value and 1-σ uncertainty of $\dot{G}/G$ from Lunar Laser Ranging (LLR). Generally only the *upper half* of the diagram has physical meaning, as the radiation of dipolar GWs must necessarily make the system lose orbital energy.

formally more accurate than our spectroscopic constraint, it might still be dominated by systematics on the distance, arising from correlations between the parallax and DM variations (see Paper II for details).

Finally, the interpretation of the mass estimates within the context of our current understanding for binary evolution implies that a significant fraction of the accreted material during the LMXB phase is ejected by the system. The discovery and study of more similar systems in the future will allow further tests of this result.

4.6 Summary of Results Presented in Paper II

Paper II reports the results of a 10-year timing campaign on PSR J1738+0333, a 5.85-ms pulsar in a low-eccentricity 8.5 h orbit with a low-mass white dwarf companion (summarized in Table 4.2). It is quite fortunate that the timing precision of PSR J1738+0333 allows a precise measurement of the key observables necessary for an estimation of the intrinsic orbital decay ($\dot{P}_b$, μ_α, μ_δ and π_x) and that the

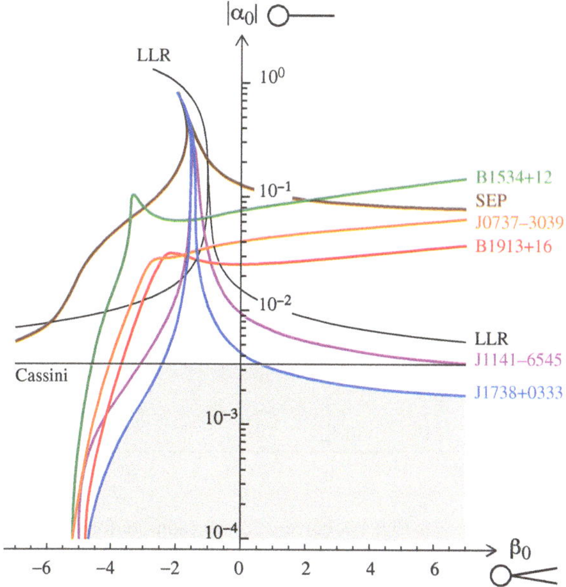

Fig. 4.7 Solar-system and binary pulsar 1-σ constraints on the matter-scalar coupling constants α_0 and β_0. Note that a logarithmic scale is used for the *vertical axis* $|\alpha_0|$, i.e., that GR ($\alpha_0 = \beta_0 = 0$) is sent at an infinite distance down this axis. LLR stands for lunar laser ranging, Cassini for the measurement of a Shapiro time-delay variation in the Solar System, and SEP for tests of the strong equivalence principle using a set of neutron star-white dwarf low-eccentricity binaries (see text). The allowed region is shaded, and it includes general relativity. PSR J1738+0333 is the most constraining binary pulsar, although the Cassini bound is still better for a finite range of quadratic coupling β_0.

optical observations provide a precise estimate of a general relativistic prediction for the orbital decay. The latter is consistent with the orbital decay from the emission of gravitational waves predicted by general relativity, $\dot{P}_b^{GR} = -27.7^{+1.5}_{-1.9} \times 10^{-15}\,\mathrm{s\,s^{-1}}$, i.e., general relativity passes the test represented by the orbital decay of this system. This agreement introduces a tight upper limit on dipolar gravitational wave emission (Fig. 4.6), a prediction of most alternative theories of gravity for asymmetric binary systems such as this. In Paper II, Freire et al. use this limit to derive the most stringent constraints ever on a wide class of gravity theories, where gravity involves a scalar field contribution (Fig. 4.7). When considering general scalar-tensor theories of gravity, our new bounds are more stringent than the best current solar-system limits over most of the parameter space, and constrain the matter-scalar coupling constant α_0^2 to be below the 10^{-5} level. For the special case of the Jordan-Fierz-Brans-Dicke theory, we obtain the one-sigma bound $\alpha_0^2 < 2 \times 10^{-5}$, which is within a factor two of the Cassini limit.

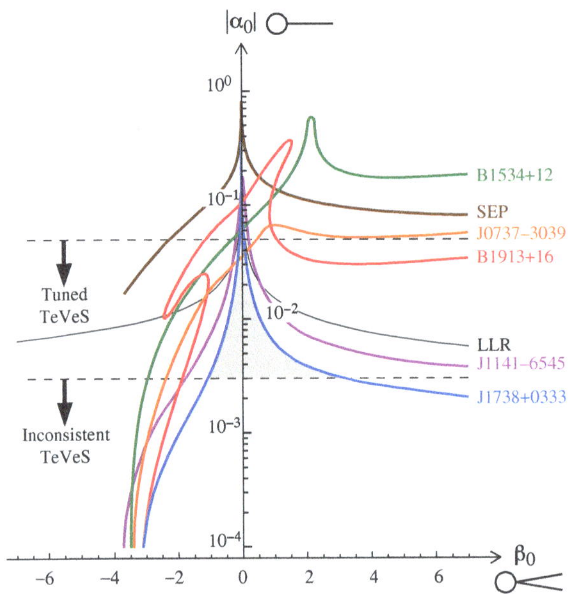

Fig. 4.8 Similar theory plane as in Fig. 4.7, but now for the (non-conformal) matter-scalar coupling described in the text, generalizing the TeVeS model. Above the *upper horizontal dashed line*, the nonlinear kinetic term of the scalar field may be a natural function; between the two *dashed lines*, this function needs to be tuned; and below the lower *dashed line*, it cannot exist any longer. The allowed region is shaded. It excludes general relativity ($\alpha_0 = \beta_0 = 0$) because such models are built to predict modified Newtonian dynamics (MOND) at large distances. Note that binary pulsars are more constraining than Solar-System tests for this class of models (and that the Cassini bound of Fig. 4.7 does not exist any longer here). For a generic nonzero β_0, PSR J1738+0333 is again the most constraining binary pulsar, while for $\beta_0 \approx 0$, the magnitude of $|\alpha_0|$ is bounded by the J0737−3039 system.

Freire et al. also use the limit on dipolar gravitational wave emission to constrain a wide class of theories of gravity which are based on a generalization of Bekenstein's Tensor-Vector-Scalar gravity (TeVeS), a relativistic formulation of Modified Newtonian Dynamics (MOND). PSR J1738+0333 is also the most constraining test of TeVeS-like theories when the quadratic matter-scalar coupling constant $|\beta_0| \geq 0.1$ (Fig. 4.8). In fact, for $\beta_0 < -1$ and $\beta_0 > 3$, such theories are excluded altogether. Bekenstein's TeVeS (a special case with $\beta_0 = 0$) is still allowed by the results of this experiment, but already needs some tuning given the small limit $|\alpha_0| < 0.035$ that we obtain from the double pulsar results (Kramer et al. 2006). We note that the precision of the latter result has greatly improved since 2006 and will be presented in a forthcoming publication (Kramer et al., in prep.). This will significantly reduce the allowed values of $|\alpha_0|$ in the gap around $\beta_0 = 0$. As a consequence, all surviving TeVeS-like theories will have to be unnaturally fine-tuned, including Bekenstein's TeVeS.

References

Bagnulo S., Jehin E., Ledoux C., Cabanac R., Melo C., Gilmozzi R., The ESO Paranal Science Operations Team 2003, The Messenger, 114, 10
Bassa C. G., van Kerkwijk M. H., Koester D., Verbunt F., 2006, A&A, 456, 295
Bessell M. S., 1990, PASP, 102, 1181
Breton R. P., Rappaport S. A., van Kerkwijk M. H., Carter J. A., 2011, ArXiv e-prints
Callanan P. J., Garnavich P. M., Koester D., 1998, MNRAS, 298, 207
Clemens J. C., Crain J. A., Anderson R., 2004, in A. F. M. Moorwood & M. Iye ed., Society of Photo-Optical Instrumentation Engineers (SPIE) Conference Series, Vol. 5492 of Society of Photo-Optical Instrumentation Engineers (SPIE) Conference Series, The Goodman spectrograph. pp 331–340
Cox A. N., 2000, Allen's astrophysical quantities
Demorest P. B., Pennucci T., Ransom S. M., Roberts M. S. E., Hessels J. W. T., 2010, Nature, 467, 1081
Drimmel R., Cabrera-Lavers A., López-Corredoira M., 2003a, A&A, 409, 205
Freire P. C. C., Bassa C. G., Wex N., Stairs I. H., Champion D. J., Ransom S. M., Lazarus P., Kaspi V. M., Hessels J. W. T., Kramer M., 2011, MNRAS, 412, 2763
Freire P. C. C., Wex N., Esposito-Farèse G., Verbiest J. P. W., Bailes M., Jacoby B. A., Kramer M., Stairs I. H., Antoniadis J., Janssen G. H., 2012, MNRAS, 423, 3328
Fukugita M., Ichikawa T., Gunn J. E., 1996, Doi M., Shimasaku K., Schneider D. P, AJ, 111, 1748
Gray D. F., 2005, The Observation and Analysis of StellarPhotospheres
Hessels J. W. T., Ransom S. M., Stairs I. H., Freire P. C. C., Kaspi V. M., Camilo F., 2006, Science, 311, 1901
Hobbs G., Lorimer D. R., Lyne A. G., Kramer M., 2005, MNRAS,360, 974
Horne K., 1986, PASP, 98, 609
Illarionov A. F., Sunyaev R. A., 1975, A&A, 39, 185
Jacoby B. A., 2005, PhD Thesis. http://adsabs.harvard.edu/abs/2005PhDT.........5J
Jacoby B. A., Bailes M., Ord S. M., Knight H. S., Hotan A. W., 2007, APJ, 656, 408
Jacoby B. A., Hotan A., Bailes M., Ord S., Kuklarni S. R., 2005, APJ, 629, L113
Kaspi V. M., Taylor J. H., Ryba M., 1994, ApJ, 428, 713
Kenyon S. J., Bromley B. C., Geller M. J., Brown W. R., 2008, APJ, 680, 312
Koester D., 2008, ArXiv: 0812.0482

Kramer M., Stairs I. H., Manchester R. N., McLaughlin M. A., Lyne A. G., Ferdman R. D., Burgay M., Lorimer D. R., Possenti A., D'Amico N., Sarkissian J. M., Reynolds J. E., Joshi B. C., Freire P. C. C., Camilo F., 2006, Science
Kuijken K., Gilmore G., 1989, MNRAS, 239, 605
Lattimer J. H., Prakash M., 2004, Science, 304, 536
Lazaridis K., Wex N., Jessner A., Kramer M., Stappers B. W., Janssen G. H., Desvignes G., Purver M. B., Cognard I., Theureau G., Lyne A. G., Jordan C. A., Zensus J. A., 2009a, MNRAS, 400, 805
Lorimer D. R., 2008, Living Reviews in Relativity, 11.
Mermilliod J.-C., Weis E. W., Duquennoy A., Mayor M., 1990, A&A, 235, 114
Nice D. J., Stairs I. H., Kasian L. E., 2008, inC. Bassa, Z. Wang, A. Cumming, & V. M. Kaspi ed., 40 Years of Pulsars: Millisecond Pulsars, Magnetars and More, Vol. 983 of American Institute of Physics Conference Series, Masses of Neutron Stars in Binary Pulsar Systems. pp 453–458
Nice D. J., Taylor J. H., 1995, ApJ, 441, 429
Oke J. B., Cohen J. G., Carr M., Cromer J., Dingizian A., Harris F. H., Labrecque S., Lucinio R., Schaal W., Epps H., Miller J., 1995, PASP, 107, 375
Paczynski B., 1990, ApJ, 348, 485
Panei J. A., Althaus L. G., Benvenuto O. G., 2000, A&A,353, 970
Panei J. A., Althaus L. G., Chen X., Han Z., 2007, MNRAS, 382, 779
Phinney E. S., 1992, Phil. Trans.:Phys. Sc. & Eng., 341, 39
Phinney E. S., Kulkarni S. R., 1994, ARA&A, 32, 591
Pylyser E., Savonije G. J., 1989, A&A, 208, 52
Reid I. N., 1996, AJ, 111, 2000
Savonije G. J., 1987, 325, 416
Schlegel D. J., Finkbeiner D. P., Davis M., 1998a, ApJ,500, 525
Serenelli A. M., Althaus L. G., Rohrmann R. D., Benvenuto O. G., 2001, MNRAS, 325, 607
Serenelli A. M., Althaus L. G., Rohrmann R. D., Benvenuto O. G., 2002, MNRAS, 337, 1091
Shapiro I. I., 1964, Phys. Rev. Lett., 13, 789
Stetson P. B., 1990, 102, 932
Tauris T. M., 2011, ArXiv e-prints
Tauris T. M., Bailes M., 1996, A&A, 315
Tauris T. M., Savonije G. J., 1999, A& A, 350, 928
Taylor J. H., Weisberg J. M., 1982, ApJ, 253, 908
Tremblay P.-E., Bergeron P., 2009, ApJ, 696, 1755
van Kerkwijk M. H., Bergeron P., Kulkarni S. R., 1996, ApJ, 467, L89
van Kerkwijk M. H., Rappaport S. A., Breton R. P., Justham S., Podsiadlowski P., Han Z., 2010b, APJ, 715, 51
Webbink R. F., Rappaport S., Savonije G. J., 1983, ApJ, 270, 678
Weisberg J. M., Nice D. J., Taylor J. H., 2010, ApJ, 722, 1030
Zwitter T., Castelli F., Munari U., 2004, A&A, 417, 1055

Chapter 5
A Massive Pulsar in a Compact Relativistic Binary

Abstract Many physically motivated extensions to general relativity (GR) predict significant deviations at energies present in massive neutron stars. We report the measurement of a 2.01 ± 0.04 solar mass ($M_{\odot}$) pulsar in a 2.46-h orbit around a $0.172 \pm 0.003\ M_{\odot}$ white dwarf. The high pulsar mass and the compact orbit make this system a sensitive laboratory of a previously untested strong-field gravity regime. Thus far, the observed orbital decay agrees with GR, supporting its validity even for the extreme conditions present in the system. The resulting constraints on deviations support the use of GR-based templates for ground-based gravitational wave detection experiments. Additionally, the system strengthens recent constraints on the properties of dense matter and provides novel insight to binary stellar astrophysics and pulsar recycling.

Massive neutron stars (NSs; $M > 1.8\,M_{\odot}$) observed as radio pulsars probe fundamental physical laws in conditions unique in the observable Universe and inaccessible to terrestrial experiments. Their high masses, directly linked to the equation-of-state (EOS) of matter at supra-nuclear densities (Lattimer and Prakash 2004; Demorest et al. 2010), constrain the lower mass limit for production of astrophysical black holes (BHs). Furthermore, they possess extreme gravitational fields which results in gravitational binding energies significantly higher than those found in regular, $1.4\,M_{\odot}$ NSs. Modifications to GR, often motivated by the desire for a unified model of the four fundamental forces, can generally imprint measurable signatures in gravitational waves (GWs) radiated by systems containing such objects, even if deviations from GR vanish in the Solar System and in lower-mass NSs (Damour and Esposito-Farese 1993, 1996; Will 1993). However, the most massive NSs known today reside in much longer-period binaries or other systems unsuitable for GW radiation tests. Identifying a massive NS in a compact, relativistic binary is thus of key importance for understanding gravity-matter coupling under extreme conditions.

Among the discoveries of a recent survey (Boyles et al. 2012; Lynch et al. 2012) conducted with the Green Bank Telescope (GBT) was PSRJ0348+0432, a pulsar spinning at 39 ms in a 2.46-h orbit around a low-mass companion. Initial timing observations of the binary yielded an accurate astrometric position, which

Summary of results presented in Antoniadis et al. 2013, http://adsabs.harvard.edu/abs/2013Sci...340..448A.

J. Antoniadis, *Multi-Wavelength Studies of Pulsars and Their Companions*, Springer Theses, DOI 10.1007/978-3-319-09897-5_5

allowed us to identify its optical counterpart in the Sloan Digital Sky Survey archive. The colors and flux of the counterpart are consistent with a low-mass white dwarf (WD) with a helium core at a distance of $d \sim 2.1$ kpc. Its relatively high brightness ($g' = 20.71 \pm 0.03$ mag) allowed us to resolve its spectrum using the Apache Point Optical Telescope. These observations revealed deep Hydrogen lines, typical of low-mass WDs, confirming our preliminary identification. The radial velocity of the WD mirrored that of PSR J0348+0432, also verifying that they are gravitationally bound. In December 2011 we obtained phase-resolved spectra of the optical counterpart using the FORS2 spectrograph of the Very Large Telescope (VLT). For each spectrum, we measured the radial velocity which we then folded modulo the system's orbital period. Our orbital fit to the velocities constrains the semi-amplitude of their modulation to be $K_{\rm WD} = 351 \pm 4\,{\rm km\,s^{-1}}$ (Fig. 5.1). Similarly, the orbital solution from radio-pulsar timing yields $K_{\rm PSR} = 30.008235 \pm 0.000016\,{\rm km\,s^{-1}}$ for the pulsar. Combined, these constraints imply a mass ratio, $q = M_{\rm PSR}/M_{\rm WD} = K_{\rm WD}/K_{\rm PSR} = 11.70 \pm 0.13$. Modeling of the Balmer-series lines in a high signal-to-noise average spectrum formed by the coherent addition of individual spectra shows that the WD has an effective temperature of $T_{\rm eff} = (10120 \pm 47_{\rm stat} \pm 90_{\rm sys})$ K and a surface gravity of $\log(g\,[{\rm cm\,s^{-2}}]) = (6.035 \pm 0.032_{\rm stat} \pm 0.060_{\rm sys})$ dex. Here each uncertainty represents an estimate of the statistical and systematic uncertainties of our fit. We found no correlation of this measurement with orbital phase and no signs of rotationally-induced broadening in the spectral lines.

The surface gravity of the WD scales with its mass and the inverse square of its radius ($g \equiv GM_{\rm WD}/R_{\rm WD}^2$). Thus, the observational constraints combined with a theoretical finite temperature mass-radius relation for low-mass WDs yield a unique solution for the mass of the companion (van Kerkwijk et al. 2005). Considering modeling uncertainties, such as the metallicity of the progenitor star and reduction of the envelope's thickness due to post-collapse hydrogen shell flashes, we find that the observed spectrum is consistent with a WD mass in the range $0.165 - 0.185\,M_\odot$ at 99.73 % confidence. These constraints are shown in Fig. 5.1. The derived WD mass and mass ratio q imply a NS mass in the range $1.93 - 2.12\,M_\odot$ at 95.45 % or $1.90 - 2.18\,M_\odot$ at 99.73 % confidence. Hence, PSR J0348+0432 is only the second NS with a precisely determined mass close to $2\,M_\odot$, after PSR J1614−2230 (Demorest et al. 2010). It has a 3σ lower mass limit $0.05\,M_\odot$ higher than the latter, and therefore provides a verification, using a different method, of the constraints on the EOS of super-dense matter present in NS interiors (Özel et al. 2010; Demorest et al. 2010). Now, for these masses and the known orbital period, GR predicts that the orbital period should decrease at the rate of $\dot{P}_{\rm b}^{\rm GR} = (-2.58^{+0.07}_{-0.11}) \times 10^{-13}\,{\rm s\,s^{-1}}$ (68.27 % C.L.) due to energy loss through GW emission.

Since April 2011 we have been observing PSR J0348+0432 with the 1.4 GHz receiver of the 305-m radio telescope at the Arecibo Observatory, using its four Wide-band Pulsar Processors. In order to verify the Arecibo data, we have been independently timing PSR J0348+0432 at 1.4 GHz using the 100-m radio telescope in Effelsberg, Germany. The timing data sets produce consistent rotational models, providing added confidence in both. Combining the Arecibo and

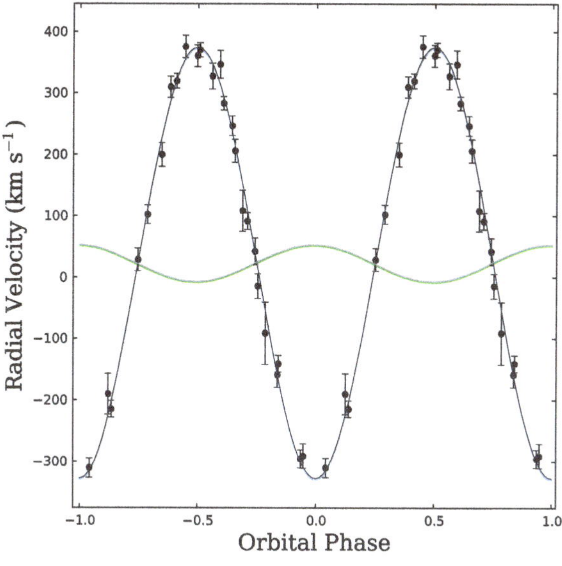

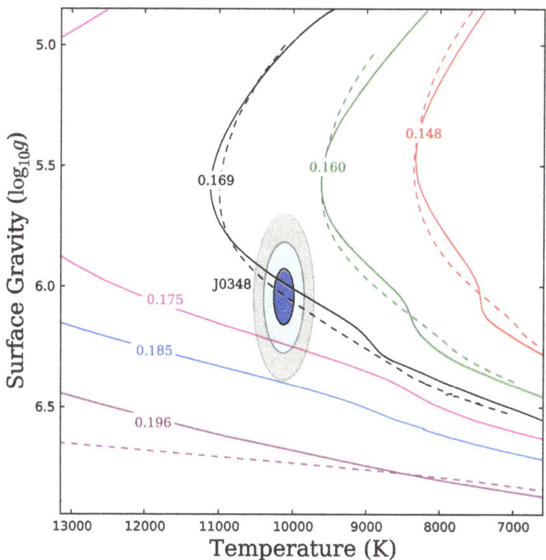

Fig. 5.1 *Upper* Radial velocities of the WD companion to PSR J0348+0432 plotted against the orbital phase (shown twice for clarity). Over-plotted is the best-fit orbit of the WD (*blue line*) and the mirror orbit of the pulsar (*green*). *Lower* Constraints on effective temperature, $T_{\rm eff}$, and surface gravity, g, for the WD companion to PSR J0348+0432 compared with theoretical WD models. The *shaded areas* depict the $\chi^2 - \chi^2_{\rm min} = 2.3, 6.2$ and 11.8 intervals (equivalent to 1, 2 and 3 σ) of our fit to the average spectrum. *Dashed lines* show the detailed theoretical cooling models of (Serenelli et al. 2001). *Continuous lines* depict tracks with thick envelopes for masses up to $\sim 0.2\,{\rm M}_\odot$ that yield the most conservative constraints for the mass of the WD.

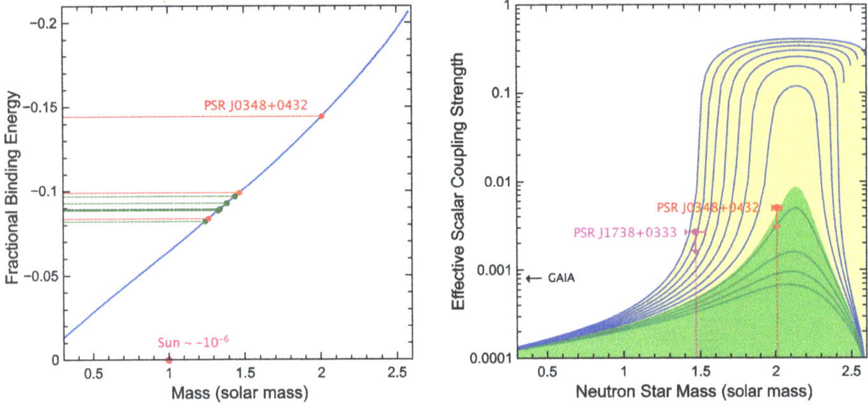

Fig. 5.2 *Left* Fractional gravitational binding energy as a function of the inertial mass of a NS in GR (*blue curve*). The *dots* indicate the NSs of relativistic NS-NS (in *green*) and NS-WD (in *red*) binary-pulsar systems currently used for precision gravity tests. *Right* Effective scalar coupling as a function of the NS mass, in the "quadratic" ST theory of (Damour and Esposito-Farese 1996). For the linear coupling of matter to the scalar field we have chosen $\alpha_0 = 10^{-4}$, a value well below the sensitivity of any near-future Solar System experiment [e.g. GAIA (Hobbs et al. 2010)]. The *solid curves* correspond to stable NS configurations for different values of the quadratic coupling β_0: -5 to -4 (*top* to *bottom*) in steps of 0.1. The *yellow area* indicates the parameter space allowed by the best current limit on $|\alpha_{\rm PSR} - \alpha_0|$ (Freire et al. 2012), while only the *green area* is in agreement with the limit presented here. PSR J0348+0432 probes deeper into the non-linear strong-field regime due to its high mass. The specific theory and EOS have been chosen for demonstration purposes.

Effelsberg data with the initial GBT observations from (Lynch et al. 2012), we derive a timing solution that includes a significant measurement of orbital decay, $\dot{P}_{\rm b} = (-2.73 \pm 0.45) \times 10^{-13}\,{\rm s\,s^{-1}}$ (68.27 % C.L.). We find that the kinematic corrections from relative motion in the Galaxy are negligible; hence this $\dot{P}_{\rm b}$ is intrinsic to the system. Furthermore, we can exclude significant contributions to $\dot{P}_{\rm b}$ from tidal effects. Therefore, the observed $\dot{P}_{\rm b}$ is caused by GW emission and its magnitude is entirely consistent with the GR prediction: $\dot{P}_{\rm b}/\dot{P}_{\rm b}^{\rm GR} = 1.05 \pm 0.18$.

Due to its large mass, PSR J0348+0432 has a gravitational binding energy ~50 % higher than the best-studied NSs in relativistic systems (see Fig. 5.2). As the magnitude of strong-field effects generally depends non-linearly on the binding energy, the measurement of orbital decay transforms the system into a gravitational laboratory for a yet untested regime, qualitatively very different from what was accesible in the past. There are compelling arguments that GR is not valid beyond a (yet unknown) critical point, like its incompatibility with quantum theory and its prediction of singularities under generic conditions. In physically consistent and extensively studied alternatives, gravity is generally mediated by extra fields (e.g. scalar) in addition to the tensor field of GR (Will 1993). A dynamical coupling between matter and these extra fields can lead to prominent deviations from GR that only occur at the high gravitational binding energies of massive NSs (Damour and Esposito-Farese 1993, 1996). In the PSR J0348+0432 system, where such an object is closely orbited by a weakly self-gravitating body, one generally expects a violation of the strong equivalence principle that in turn leads to emission of dipolar gravitational radiation (Will 1993). The leading change in the orbital period is then:

$$\dot{P}_{\mathrm{b}}^{\mathrm{dipolar}} \simeq -\frac{4\pi^2 G}{c^3 P_{\mathrm{b}}} \frac{M_{\mathrm{PSR}} M_{\mathrm{WD}}}{M_{\mathrm{PSR}} + M_{\mathrm{WD}}} \left(\alpha_{\mathrm{PSR}} - \alpha_{\mathrm{WD}}\right)^2 , \tag{5.1}$$

where α_{PSR} is the effective coupling strength between the NS and the ambient fields responsible for the dipole moment [e.g. scalar fields in scalar-tensor (ST) gravity] and α_{WD} is the same parameter for the WD which, given its weak field is practically identical to the linear coupling α_0 that is well constrained in the Solar System (Will 1993). A significant α_{PSR} for NSs up to $1.47\,\mathrm{M}_\odot$ has been excluded by various binary pulsar experiments (e.g. Freire et al. 2012). The consistency of GW emission with the predictions of GR for PSR J0348+0432 implies $|\alpha_{\mathrm{PSR}} - \alpha_0| < 0.005$ (95 % C.L.), and consequently excludes significant strong-field deviations, even for massive NSs of $\sim 2\,\mathrm{M}_\odot$. Figure 5.2 shows this result in the context of ST gravity.

This limit on a deviation from GR also has important consequences for ground-based GW experiments, like LIGO and VIRGO. Their most promising sources are in-spiralling compact binaries, consisting of NSs or BHs (Sathyaprakash and Schutz 2009). The GW signal from such systems will be deeply buried in the instrumental noise. Thus, detection will require correlation of the signal with detailed model templates, currently constructed within the GR framework (Sathyaprakash and Schutz 2009). However, dipolar gravitational radiation would lead to a modified phase evolution of the inspiral, which might strongly deviate from the GR prediction (Will 1994; Damour and Esposito-Farese 1998). With the results presented here, we can now exclude a deviation of more than 0.5 cycles from the GR template during the observable in-spiral caused by additional long-range gravitational fields, for the whole range of NS masses observed in nature. Furthermore, in an extension of the arguments in (Will 1994; Damour and Esposito-Farese 1998), our result implies that binary pulsar experiments are already more sensitive for testing such deviations than the upcoming advanced GW detectors.

The measured spin rate and spin-rate change (Lynch et al. 2012); of PSRJ0348+ 0432, combined with the masses and orbital period of the system, form a peculiar set of parameters that brings new insight to binary stellar evolution. The short orbital period of 2.46-h is best understood from evolution via a common envelope (CE) where the NS is captured in the envelope of the WD progenitor, leading to efficient removal of orbital angular momentum on a short timescale(Iben and Livio 1993). An interesting consequence of this scenario is that the NS was born with an initial mass close to its current mass, since very little accretion was possible. While the slow spin rate of $\sim$39 ms and the unusually high magnetic field of a few 10^9 G provide further support, the low WD mass contradicts the standard CE hypothesis: it requires a progenitor star mass $M_2 < 2.2\,\mathrm{M}_\odot$, since more massive stars would leave behind much more massive cores. For such low donor star masses, the mass ratio of the binary components is close to unity, leading to dynamically stable mass transfer without forming a CE (Podsiadlowski et al. 2002). A speculative hypothesis to circumvent this problem would be a CE evolution with hypercritical accretion, where $\sim 0.6\,\mathrm{M}_\odot$ of material was efficiently transferred to a $1.4\,\mathrm{M}_\odot$ NS (Chevalier 1993).

An alternative formation scenario is evolution via a close-orbit low-mass X-ray binary with a 1.0–1.6 $M_{\odot}$ donor star that suffered from loss of orbital angular momentum due to magnetic braking (Pylyser and Savonije 1989). This requires a finely tuned truncation of the mass-transfer process which is not yet understood in detail, but is also required for other known recycled pulsars with short orbital periods and WD companions. The slow spin rate could then be explained from a combination of spin-down during the Roche-lobe decoupling phase (Tauris 2012) and subsequent magnetic dipole radiation from the high-magnetic-field pulsar.

Emission of GWs will continue to shrink the orbit of PSR J0348+0432 and in 400 Myr (when $P_{\rm b} \simeq 23$ min) the WD will fill its Roche-lobe and possibly leave behind a planet orbiting the pulsar (Bailes et al. 2011), unless a BH forms via accretion-induced collapse of the massive NS in a cataclysmic, γ-ray burst-like event (Dermer and Atoyan 2006).

References

Bailes M., Bates S. D., Bhalerao V., Bhat N. D. R., Burgay M., Burke-Spolaor S., D'Amico N., Johnston S., Keith M. J., Kramer M., Kulkarni S. R., Levin L., Lyne A. G., Milia S., Possenti A., Spitler L., Stappers B., van Straten W., 2011, Science, 333, 1717.

Boyles J., Lynch R. S., Ransom S. M., Stairs I. H., Lorimer D. R., McLaughlin M. A., Hessels J. W. T., Kaspi V. M., Kondratiev V. I., Archibald A., Berndsen A., Cardoso R. F., Cherry A., Epstein C. R., Karako-Argaman C., McPhee C. A., 2012. ArXiv:1209.4293.

Chevalier R. A., 1993, ApJ, 411, L33.

Damour T., Esposito-Farese G., 1993, Phys. Rev. Lett., 70, 2220.

Damour T., Esposito-Farese G., 1996, Phys. Rev. D, 54, 1474.

Damour T., Esposito-Farese G., 1998, Phys. Rev. D, 58, 1.

Demorest P. B., Pennucci T., Ransom S. M., Roberts M. S. E., Hessels J. W. T., 2010, Nature, 467, 1081.

Dermer C. D., Atoyan A., 2006, ApJ, 643, L13.

Freire P. C. C., Wex N., Esposito-Farèse G., Verbiest J. P. W., Bailes M., Jacoby B. A., Kramer M., Stairs I. H., Antoniadis J., Janssen G. H., 2012, MNRAS, 423, 3328.

Hobbs D., Holl B., Lindegren L., Raison F., Klioner S., Butkevich A., 2010, in Klioner S. A., Seidelmann P. K., Soffel M. H., eds, IAU Symposium, Vol. 261 of IAU Symposium, Determining PPN γ with Gaia's astrometric core solution. pp 315–319.

Iben Jr. I., Livio M., 1993, PASP, 105, 1373.

Lattimer J. H., Prakash M., 2004, Science, 304, 536.

Lynch R. S., Boyles J., Ransom S. M., Stairs I. H., Lorimer D. R., McLaughlin M. A., Hessels J. W. T., Kaspi V. M., Kondratiev V. I., Archibald A. M., Berndsen A., Cardoso R. F., Cherry A., Epstein C. R., 2012. ArXiv:1209.4296.

Özel F., Psaltis D., Ransom S., Demorest P., Alford M., 2010, ApJ, 724, L199.

Podsiadlowski P., Rappaport S., Pfahl E. D., 2002, ApJ, 565, 1107.

Pylyser E., Savonije G. J., 1989, A&A, 208, 52.

Sathyaprakash B. S., Schutz B. F., 2009, Living Reviews in Relativity, 12, 2.

Serenelli A. M., Althaus L. G., Rohrmann R. D., Benvenuto O. G., 2001, MNRAS, 325, 607.

Tauris T. M., 2012, Science, 335, 561.

van Kerkwijk M., Bassa C. G., Jacoby B. A., Jonker P. G., 2005, Optical Studies of Companions to Millisecond Pulsars.

Will C. M., 1993, Theory and Experiment in Gravitational Physics.

Will C. M., 1994, Phys. Rev. D, 50, 6058.

Chapter 6
A White Dwarf Companion to the Relativistic Pulsar J1141−6545

Abstract Pulsars with compact companions in close eccentric orbits are unique laboratories for testing general relativity and alternative theories of gravity. Moreover, they are excellent targets for future gravitational wave experiments like LISA and they are also highly important for understanding the equation of state of super-dense matter and the evolution of massive binaries. Here we report on optical observations of the $1.02\,M_{\odot}$ companion to the pulsar PSR J1141−6545. We detect an optical counterpart with apparent magnitudes $V = 25.08(11)$ and $R = 24.38(14)$, consistent with the timing position of the pulsar. We demonstrate that our results are in agreement with a white dwarf companion. However the latter is redder than expected and the inferred values are not consistent with the theoretical cooling tracks, preventing us from deriving the exact age. Our results confirm the importance of the PSR J1141−6545 system for gravitational experiments.

6.1 Introduction

The value of relativistic binaries is highly recognised, as their study can provide insight into some of the holy grails of fundamental physics. Among them are tests of general relativity and alternative theories of gravity, the detection of gravitational waves, the study of the equation of state of super-dense matter and tests of evolutionary scenarios for heavy stars (for a complete review see Lorimer and Kramer 2004).

The sample of relativistic binaries discovered so far is dominated by double neutron stars, covering a wide range of orbital parameters. Another substantial fraction consists of white dwarf–neutron star binaries, most of them in almost perfectly circular orbits (e.g. review by van Kerkwijk et al. 2005). These systems are the result of the evolution of a massive primary which evolves fast, explodes as a supernova and becomes a neutron star (NS); and of a lighter secondary which evolves slower and eventually becomes a white dwarf (WD) (Driebe et al. 1998). During the final interaction phase, the NS is spun up to very short rotation periods and becomes

Based on Antoniadis et al. 2011, http://adsabs.harvard.edu/abs/2011MNRAS.412..580A.

J. Antoniadis, *Multi-Wavelength Studies of Pulsars and Their Companions*,
Springer Theses, DOI 10.1007/978-3-319-09897-5_6

a millisecond pulsar. Any eccentricity (primordial or resulting from the supernova kick) is dampened by tidal interaction before the secondary becomes a WD.

A significant exception to the preceding is the binary PSR B2303+46 (Stokes et al. 1985). In that system, the WD (van Kerkwijk and Kulkarni 1999) orbits the non-recycled pulsar in a highly eccentric orbit. Investigations into possible formation scenarios for this type of binary (Tauris and Sennels 2000; Davies et al. 2002; Church et al. 2006) have shown that they most likely originate from a binary system of massive stars with nearly equal mass. When the initially more massive star reaches the red giant phase, the secondary star accretes sufficient mass to surpass the Chandrasekhar limit, allowing it to eventually evolve into a NS. The primary star, however, loses sufficient mass to end up as a heavy WD. Hence, in the resulting system, the WD is expected to be older than the pulsar.

The only other promising candidate for this category is the PSR J1141−6545 binary system, initially discovered in a Parkes survey (Kaspi et al. 2000). PSR J1141−6545 is a 0.2 day binary in an eccentric orbit ($e \sim 0.17$, Bhat et al. 2008). The primary is a relatively young 394 ms pulsar (characteristic age $\sim$ 1.4 Myrs), orbited by a compact object of unknown nature. Bhat et al. (2008) derived $M_c = 1.02(1)\,M_\odot$ for the mass of the companion by applying the relativistic DDGR orbital model (Damour and Deruelle 1986) to their timing measurements. The latter is consistent with both a heavy WD and a light NS with the former case being more favoured by statistical evidence (Tauris and Sennels 2000). Jacoby et al. (2006) included the system in an optical survey but found no optical counterpart down to $R = 23.4$.

This paper reports on optical observations of the companion star in the PSR J1141−6545 binary system. Our main scientific rationale for this study is that in the case of a positive WD confirmation, the system would be of great importance for gravitational tests. In particular, because of its gravitational asymmetry, PSR J1141−6545 would be one of the most constraining systems known for general relativity in the strong field regime as it is expected to emit strong dipolar gravitational radiation in a wide range of scalar-tensor theories (Will 1993; Esposito-Farese 2005; Bhat et al. 2008).

The structure of the text is as follows: in Sect. 6.2 we describe the observations and the data reduction process while in Sect. 6.3 we present our results. Finally, in Sect. 6.4 we discuss our findings and comment on their astrophysical consequences and their importance in gravitational tests.

6.2 Observations and Data Reduction

We have obtained optical images in the V-band and R-band filters, of the field containing PSR J1141−6545 using the FORS1 instrument mounted at the UT2 of the Very Large Telescope (VLT). Both filters resemble the standard Johnson-Cousin filters but have slightly higher sensitivity in the red, sharper cut offs and higher throughput. The observations were conducted in service mode during the night of 6th of April 2008. The conditions were photometric and the average seeing of the night was 0″.7. The total exposure time was 600 s in V and 1,500 s in R. In order to minimize potential problems with cosmic rays and guiding errors and avoid saturation of bright stars, the exposures were split in three sub-exposures of 200 s in the V-band and three sub-exposures of 500 s in the R-band. For the data reduction we used the FORS1 pipeline provided by ESO. Each image was first bias corrected and flat-fielded using twilight flats. Bad pixels and cosmic ray hits in all frames were replaced by a median over their neighbourghs. The resulting frames were then sky-subtracted, registered and combined in one averaged frame for each filter.

6.2.1 Photometry

We performed point-spread function (PSF) photometry on the average frame of each filter using DAOPHOT II (Stetson 1987) inside the Munich Image Data Analysis System (MIDAS). The PSF was determined following a slightly modified version of the recipe in Stetson (1987). First, we selected 100 bright, unsaturated stars ($\leq$40,000 ADUs) located within 1′ distance from our target. Then we fitted their PSFs with a Moffat function and through an iterative process we rejected fits with root mean square (rms) residuals greater than 1 %. The stars in the vicinity of the PSF template stars were then removed with the SUBTRACT routine of DAOPHOT II and the PSF was determined again on the subtracted image, improving the rms of the fit by a factor of $\sim$2. Finally, the instrumental magnitudes of all stars within the same distance were extracted.

For the photometric calibration we first found the offset between PSF and aperture magnitudes of six isolated bright stars in both our science images. This offset was used to transform the extracted PSF magnitudes to aperture ones. Zero-points and colour terms were determined by analysing two archival images of NGC 2437 (one in each band), obtained during the 5th of April 2008. The latter contains more than 80 Stetson photometric standards (Stetson 2000). Of those, we used only 30 depicted on the same area of the CCD as our target. We determined their instrumental magnitudes using the same aperture, inner and outer sky radii as in our science images. We fitted for zero-points and colour terms using the average extinction coefficients provided by ESO (0.120(3) and 0.065(4) per airmass for V and R respectively) and used them to transform our measurements to the standard Johnson-Cousin system. The rms residual of the fit was 0.02 mag in V and 0.04 mag in R.

6.2.2 Astrometry

For the astrometric calibration we selected 58 astrometric standards from the USNO CCD Astrograph Catalogue (UCAC3, Zacharias et al. 2010) that coincided with the 7′ × 4′ averaged V image. Because of the 200 s exposure times, only 13 of them were not saturated or blended and appeared stellar. The centroids of these stars were measured and an astrometric solution, fitting for zero-point position, scale and position angle, was computed. Two outliers were iteratively removed, and the final solution, using 11 stars, had root-mean-square residuals of $0''.056$ in right ascension and $0''.058$ in declination, which is typical for the UCAC3 catalogue.

The low number of astrometric standards used makes the astrometric solution sensitive to random noise, and hence we computed another solution using stars from the 2MASS catalogue. Of the 360 stars from this catalogue that coincided with the *V* image, 245 were not saturated and appeared stellar and unblended. The iterative scheme removed outliers and converged on a solution using 210 stars with rms residuals of $0''.14$ and $0''.13$. This solution is consistent with the UCAC3 astrometric solution to within the uncertainties and we are confident in using the UCAC3 solution for the astrometric calibration.

6.3 Results

A faint star is present on the timing position of the pulsar (Manchester et al., 2010) in both the averaged *V* and *R* images (Fig. 6.1). The optical position is $\alpha_{2000} = 11^{\mathrm{h}}41^{\mathrm{m}}07^{\mathrm{s}}\!.00(2)$ and $\delta_{2000} = -65°45'19''.01(10)$, where the uncertainty is the quadratic sum of the positional uncertainty of the star (approximately $0''.08$ in both coordinates) and the uncertainty in the astrometric calibration. This position is offset from the timing position by $\Delta\alpha = -0''.08 \pm 0''.11$ in right ascension and $\Delta\delta = 0''.10 \pm 0''.12$ in declination. Hence, the timing and the optical positions agree within errors.The images have an average stellar density of 239 stars per

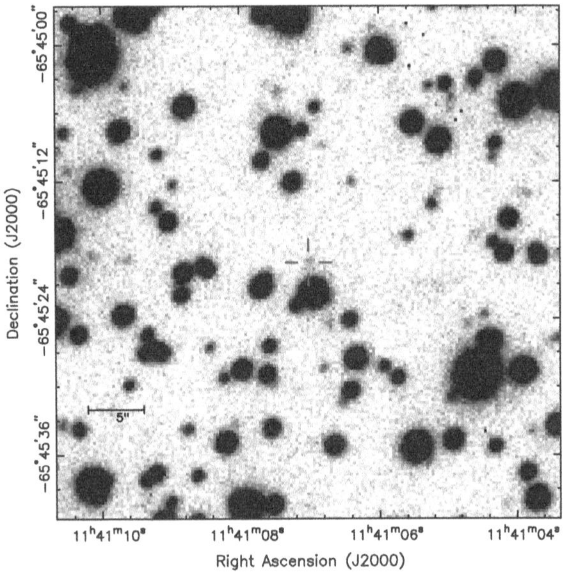

Fig. 6.1 A $45'' \times 45''$ subsection of the averaged V-band image. The timing position of PSR J1141−6545 is denoted by $1''$ tickmarks

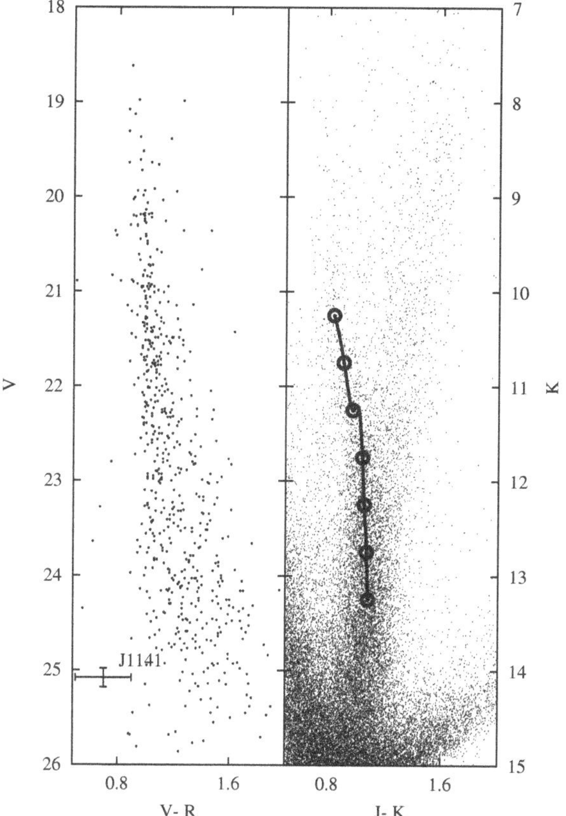

Fig. 6.2 *Left* The extracted V magnitudes of all objects in our data, plotted against their $V - R$ colours. The counterpart of PSR J1141−6545 is placed among the faintest and bluest objects. *Right* Colour-Magnitude diagram of 2MASS sources, located within 20′ distance from PSR J1141−6545. The *circles* indicate the calculated position of *red* clump stars. The *line* is a 3rd order spline connecting all *circles*

square arcminute, which translates to only a 0.9 % probability of a chance coincidence within the 95 % confidence error circle, which has a radius of $0''.20$. The star has $V = 25.08(12)$ and $R = 24.38(14)$ and at $V - R = 0.70(18)$, it is significantly bluer than the bulk of the stars in the field, which have $V - R = 1.27(30)$ for $24 < V < 26$. Any MS or post-MS star would be brighter and/or redder given the distance ($\gtrsim$3.7 kpc, Sect. 6.3), hence we are confident that the star inside the error circle is the white dwarf companion to PSR J1141−6545 (Fig. 6.2).

6.3.1 Distance and Reddening

The intrinsic color and brightness of the WD and hence its cooling age and temperature, can be inferred from our measurements under the condition of an accurate distance and reddening estimate. Unfortunately, as for most pulsars, the distance to the PSR J1141−6545 system is not well known.

An estimate can be made from the observed dispersion measure (DM) and a model of the free electron distribution in the Galaxy. Using the NE2001 Galactic free electron model (Cordes and Lazio 2002), we find $d = 2.4$ kpc for the observed DM $=$ 116.08 cm^{-3}pc (Manchester et al. 2010) towards PSR J1141−6545. Traditionally the uncertainty on DM derived distances is quoted at 20 %, however, a comparison with pulsar parallaxes indicate that the uncertainties may be as large as 60 % (Deller et al. 2009). Ord et al. (2002) placed a lower bound on the distance by measuring the HI absorption spectrum of the pulsar. They concluded that the binary must be located beyond the tangent point predicted by the Galactic rotation model of Fich et al. (1989) to be at 3.7 kpc.

The interstellar extinction towards PSR J1141−6545 was traced using the red clump stars method described in Durant and van Kerkwijk (2006). We used a sample of 44,168 stars from the 2MASS catalogue, situated within 20′ distance from PSR J1141−6545 (right panel of Fig. 6.2). We then split the sample in seven 0.5 mag-wide stripes, ranging from $K = 10$ to $K = 13.5$ and traced the $J - K$ location of the helium−core giants by fitting their distribution with a power law plus a Gaussian, as in Durant and van Kerkwijk (2006) (right panel of Fig. 6.2). We used $K_0 = -1.65$ for the intrinsic luminosity, $(J - K)_0 = 0.75$ for the intrinsic colour (Wainscoat et al. 1992; Hammersley et al. 2000) and $A_K = 0.112 A_V$ (Schlegel et al. 1998). The extinction was found to range from $A_V = 0.55$ to $A_V = 2.04$ for distances of $1.1 - 4.5$ kpc. For the 3.7 kpc distance of Ord et al. (2002), we deduce $A_V = 2.00$. Our values are smaller than the ones derived by the model of Drimmel et al. (2003) (e.g. $A_V = 2.52$ for 3.7 kpc), most likely due to the better resolution of our method, but consistent with the values derived by Marshall et al. (2006) (e.g. $A_V = 2.05$ for 3.7 kpc).

6.3.2 Age and Temperature

The thermodynamics of WDs are simple in nature, making the cooling rates and ages easy to calculate. Several models exist for a wide variety of masses and compositions (e.g. Holberg et al. 2008; Bergeron et al. 1995). In the high mass domain, the colours and temperatures derived by these models are in good agreement, independently of the chemical composition, especially for ages smaller than 8 Gyrs. Once the mass and absolute magnitudes are known, one can correlate them with a theoretical cooling track and derive the age. In the case of the PSR J1141−6545 binary, this calculation is complicated by the uncertain distance estimate and by the fact that the measured $V - R$ color is redder than expected. In order to find the age of the WD we used the O/Ne-core 1.06 $M_\odot$ cooling track of Holberg et al. (2008) and searched for the best solution in the $\{d, A_V, T_{\rm WD}\}$ parameter space by minimising the quantity:

$$\chi^2 = \frac{[V_0(d, A_V) - V_{\rm WD}(T)]^2}{\sigma_V^2} + \frac{[R_0(d, A_R) - R_{\rm WD}(T)]^2}{\sigma_R^2} \tag{6.1}$$

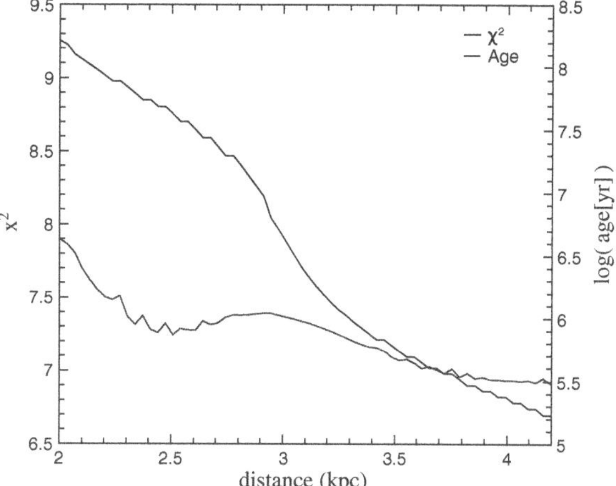

Fig. 6.3 χ^2 index (Eq. 6.1, *left axis*) and best-fit age (*right axis*) as a function of distance. The goodness of the fit continually increases with distance. The best solution is found at 4.2 kpc where the V magnitude of the WD becomes equal to the brightest value provided by the model

with V_0 and R_0 the absolute magnitudes for a given distance and reddening; $V_{\rm WD}$ and $R_{\rm WD}$ the predicted magnitudes for a given age T; and $\sigma_{\rm V,R}$ the photometric uncertainties. We varied the distance between 2 and 4.2 kpc with a 0.1 step size. For each distance, A_V was derived from our reddening calculations. Finally, the extinction was converted using $A_R = 0.819 A_V$ (Schlegel et al. 1998). Unfortunately, our method yielded no compelling solution (Fig. 6.3), not only for the most reliable $1.06\,M_\odot$ track but for other Holberg et al. (2008) and Bergeron et al. (1995) tracks of similar masses as well. In each case the minimum χ^2 was constrained by the minimum age provided by the particular model. The impact of the results on formation scenarios of PSR J1141−6545 is discussed in the next section (Fig. 6.4).

6.4 Conclusions and Discussion

The results of this paper, for the first time, provide indisputable evidence for the gravitational asymmetry of the PSR J1141−6545 binary system, i.e. its composition of a strongly self-gravitating body, the pulsar ($E^{\rm grav}/mc^2 \sim 0.2$), and a weakly self-gravitating body, the white dwarf ($E^{\rm grav}/mc^2 \sim 10^{-4}$). This is of utmost importance for testing alternative theories of gravity with this system, in particular tests of gravitational dipolar radiation. In fact, the direct observation of the white dwarf companion to PSR J1141−6545 substantiates limits on alternative gravity theories derived in the past, like in Esposito-Farese (2005), Bhat et al. (2008). Before the optical detection of the companion to PSR J1141−6545, its WD nature was inferred from the mass measurement, which is based on general relativity, and Monte-Carlo simulations of

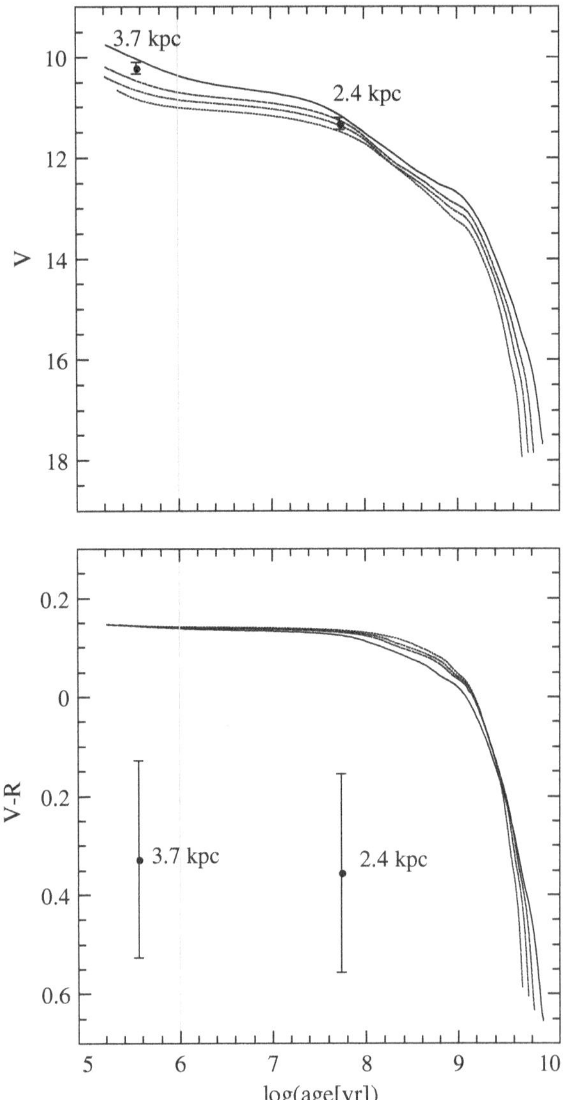

Fig. 6.4 Cooling track of a $M = 1.06\,M_\odot$ O/Ne WD (*solid line*) based on the work of Holberg et al. (2008) as reflected in its $(V - R)_0$ colour (*lower panel*) and brightness (*upper panel*). Further WD model sequences are overplotted for comparison (masses: 1.16, 1.20, $1.24\,M_\odot$; *dashed*, *dashed-dotted* and *dotted line* respectively). The color and brightness of PSR J1141−6545 is also plotted against age, for the distance of 3.7 kpc (Ord et al. 2002) and the 2.4 kpc DM distance. The error-bars in both panels represent the 1σ uncertainties derived from monte-carlo simulations of photometric and calibration errors propagation. The *grey vertical line* shows the characteristic age of the pulsar

interacting binaries. These arguments are clearly less compeling than the evidence provided here, and become debatable when testing alternative theories of gravity, in particular when performing generic tests like in the double pulsar (Kramer and Wex 2009) and the PSR J1012+5307 system (Lazaridis et al. 2009) .

References

Bergeron P., Wesemael F., Beauchamp A., 1995, PASP, 107, 1047.
Bhat N. D. R., Bailes M., Verbiest J. P. W., 2008, Phys. Rev. D, 77, 124017.
Church R. P., Bush S. J., Tout C. A., Davies M. B., 2006, MNRAS, 372, 715.
Cordes J. M., Lazio T. J. W., 2002, http://arxiv.org/abs/astro-ph/0207156
Damour T., Deruelle N., 1986, Ann. Inst. Henri Poincaré Phys. Théor., Vol. 44, No. 3, p. 263–292, 44, 263.
Davies M. B., Ritter H., King A., 2002, MNRAS, 335, 369.
Deller A. T., Tingay S. J., Bailes M., Reynolds J. E., 2009, ApJ, 701, 1243.
Driebe T., Schoenberner D., Bloecker T., Herwig F., 1998, A&A, 339, 123.
Drimmel R., Cabrera-Lavers A., López-Corredoira M., 2003, A&A, 409, 205.
Durant M., van Kerkwijk M. H., 2006, ApJ, 650, 1070.
Esposito-Farese G., 2005, in M. Novello, S. Perez Bergliaffa, & R. Ruffini ed., The Tenth Marcel Grossmann Meeting. On recent developments in theoretical and experimental general relativity, gravitation and relativistic field theories Binary-Pulsar Tests of Strong-Field Gravity and Gravitational Radiation Damping. p. 647.
Fich M., Blitz L., Stark A. A., 1989, ApJ, 342, 272.
Hammersley P. L., Garzón F., Mahoney T. J., López-Corredoira M., Torres M. A. P., 2000, MNRAS, 317, L45.
Holberg J. B., Bergeron P., Gianninas A., 2008, AJ, 135, 1239.
Jacoby B. A., Chakrabarty D., van Kerkwijk M. H., Kulkarni S. R., Kaplan D. L., 2006, ApJ, 640, L183.
Kaspi V. M., Lyne A. G., Manchester R. N., Crawford F., Camilo F., Bell J. F., D'Amico N., Stairs I. H., McKay N. P. F., Morris D. J., Possenti A., 2000, ApJ, 543, 321.
Kramer M., Wex N., 2009, Classical and Quantum Gravity, 26, 073001.
Lazaridis K., Wex N., Jessner A., Kramer M., Stappers B. W., Janssen G. H., Desvignes G., Purver M. B., Cognard I., Theureau G., Lyne A. G., Jordan C. A., Zensus J. A., 2009, MNRAS, 400, 805.
Lorimer D. R., Kramer M., 2004, Handbook of Pulsar Astronomy.
Manchester R. N., Kramer M., Stairs I. H., Burgay M., Camilo F., Hobbs G. B., Lorimer D. R., Lyne A. G., McLaughlin M. A., McPhee C. A., Possenti A., Reynolds J. E., van Straten W., 2010, ApJ, 710, 1694.
Marshall D. J., Robin A. C., Reylé C., Schultheis M., Picaud S., 2006, A&A, 453, 635.
Ord S. M., Bailes M., van Straten W., 2002, MNRAS, 337, 409.
Schlegel D. J., Finkbeiner D. P., Davis M., 1998, ApJ, 500, 525.
Stetson P. B., 1987, PASP, 99, 191.
Stetson P. B., 2000, PASP, 112, 925.
Stokes G. H., Taylor J. H., Dewey R. J., 1985, ApJ, 294, L21.
Tauris T. M., Sennels T., 2000, A&A, 355, 236.
van Kerkwijk M. H., Kulkarni S. R., 1999, ApJ, 516, L25.
van Kerkwijk M. H., Bassa C. G., Jacoby B. A., Jonker P. G., 2005, in F. A. Rasio & I. H. Stairs ed., Binary Radio Pulsars, Vol. 328 of Astronomical Society of the Pacific Conference Series, Optical Studies of Companions to Millisecond Pulsars. p. 357.
Wainscoat R. J., Cohen M., Volk K., Walker H. J., Schwartz D. E., 1992, ApJs, 83, 111.
Will C. M., 1993, Theory and Experiment in Gravitational Physics.
Zacharias N., Finch C., Girard T., Hambly N., Wycoff G., Zacharias M. I., Castillo 2010, AJ, 139, 2184.

Chapter 7
Summary and Future Work

7.1 Overview

This thesis dealt with a diverse ensemble of pulsars with white-dwarf companions. The results drawn, have ramifications for a large range of (astro)physical applications. Bellow I give some final remarks and comment on potential future perspectives.

First, in Chap. 3 we presented an analysis of the spectrum and luminosity of the white-dwarf companion to the millisecond pulsar PSR J1909−3744. Together with the Shapiro delay and parallax constraints, our observations allowed us to derive, for the first time, a model-independent measurement of the white dwarf's surface gravity and compare it with the outcome of the spectral modelling. Despite the relatively poor quality of the Gemini spectra, the qualitative comparison shows clear evidence for the "high log g" problem, first identified in higher- mass, C/O–core white dwarfs. During the course of this work, the problem was conclusively linked with the non-ideal treatment of convective energy transport in 1D atmospheric models (Tremblay et al. 2011, 2013). Until detailed 3D atmospheric grids become available, PSR J1909−3744 provides an empirical rule for correcting spectral observations of similar convective stars.

On the contrary, mass-radius relations for He-core white dwarfs with *thick* hydrogen envelopes seem to reproduce the observational constraints on PSR J1909−3744 on spot. To begin with, this is particularly reassuring for the masses of white dwarfs derived with the same technique. However, amusingly, none of the modern cooling numerical codes that include a detailed treatment of element diffusion predict the existence of such objects: for most models, a last CNO-induced hydrogen flash destroys the thick envelope of the (pre-)white dwarf if its mass is above $0.17\,M_{\odot}$. Fortunately, systems with similar masses (e.g. PSR J1738+0333) are rather insensitive to the (non)treatment of diffusion and thus older mass-radius relations can be used with safety.

Having established the regime of validity for white dwarf models, we moved on to apply the same technique on the relativistic binary PSR J1738+0333 (Chap. 4). Fortunately, due to the relatively lower mass and higher temperature, the companion's atmosphere is mostly radiative and thus the spectral constraints on surface gravity should be reliable within errors. The inferred pulsar mass, $M_{\rm NS} = 1.46 \pm 0.06\,{\rm M}_{\odot}$ is surprising low. This suggests that a large fraction of the companion's matter escaped

J. Antoniadis, *Multi-Wavelength Studies of Pulsars and Their Companions*,
Springer Theses, DOI 10.1007/978-3-319-09897-5_7

the system during the X-ray binary phase, or, alternatively, that the neutron star formed via the accretion induced collapse of a heavy O/Ne/Mg white dwarf.

Furthermore, the mass determination, together with the measurement of orbital decay transform PSR J1738+0333 in an unprecedented laboratory for strong-field gravity. The imposed constraints on dipole gravitational radiation rule out a large range of alternative theories of gravity that involve a scalar-field contribution. For most of the parameter space, these limits are better than the current Solar-system constraints; for the special case of Jordan-Brans-Dicke Scalar-Tensor gravity, the limit is only $\sim$2 times weaker than the Cassini bound, but will improve in the future, as the timing precision increases. In addition, PSR J1738+0333 sets stringent limits on a wide class of theories based on Tensor-Vector-Scalar gravity, a relativistic formulation of modified Newtonian dynamics (MOND). If one accepts that General Relativity is indeed the correct theory of gravity, then the orbital decay and mass ratio can be used to infer the component masses, and thereby use them to improve the constraints on white dwarf cooling models.

In Chap. 5 we presented optical and radio-timing observations of the newly discovered compact relativistic binary PSR J0348+0432. In this system, the white dwarf has a slightly lower surface gravity and is significantly hotter. On the one hand, this means that its atmosphere is purely radiative and therefore the spectral modelling should be completely reliable. On the other hand—contrary to PSR J1909−3744 and PSR J1738+0333—its parameters make it sensitive to the treatment of diffusion. For that reason we calculated a new set of models that take into account the remaining modelling uncertainties. Based on these *conservative* tracks we find that the neutron star has a mass of $2.01 \pm 0.03\,M_{\odot}$. This result, when considered together with the orbital decay, has interesting consequences for several disciplines of astrophysics and fundamental physics.

To begin with, PSR J0348+0432 is the astronomical object with the highest confirmed binding energy, almost 50 % higher than previously known neutron stars in relativistic binaries. This could be the source of significant deviations from General Relativity, which could not be tested by previous experiments, using pulsars with lower masses. Indeed, as the magnitude of such deviations depends non-linearly on the binding energy, the tests possible now are qualitatively very different from what was available in the past. Hence, the short orbit of the binary allows for the first time a precise test for these strong-field effects through the measurement of the binary's orbital decay. Our observations are perfectly consistent with GR and thus (a) strongly support its validity even for such extreme gravity-matter couplings and (b) already rule out strong-field phenomena predicted by extensively studied and physically motivated alternative theories.

The confirmation of GR in this regime supports the use of GR-based templates for the detection of gravitational waves from double neutron star and neutron star–black hole coalescence events with advanced ground-based detectors, like LIGO and VIRGO. These theoretical templates are the result of decades of mathematical research in GR. With the PSR J0348+0432 system, their theoretical foundation is now verified for the whole range of neutron star masses observed in nature.

The pulsar is only the second with a precisely determined mass close to 2 solar masses. Thus it independently confirms, using an entirely different methodology, the existence of such massive neutron stars, first shown by Demorest et al. (2010). In the same regard, PSR J0348+0432, with its slightly higher mass, introduces the most stringent constraints ever on the properties of cold matter at supra-nuclear densities.

Last but not least, the system has a peculiar combination of properties and poses a challenge to our understanding of binary evolution and pulsar recycling. Furthermore, due to its short merging timescale of only 400 Myr, it provides t a direct link to the formation of an ultra-compact X-ray binary within a Hubble-time, possibly leading to a pulsar–planet system like PSR J1719−1438 (Bailes et al. 2011), or a short γ-ray burst in case its mass is near the upper neutron star mass limit.

Finally, in Chap. 6 I presented photometric observations of the binary PSR J1141−6545. For that system, the combination of high eccentricity and short orbital period allow the measurement of a range of relativistic effects including the orbital decay, Einstein and Shapiro delays. From these effects the companion mass is constrained to $\sim$1 $M_{\odot}$ which is consistent with both a heavy white dwarf and a light neutron star formed through an electron-capture supernova. Our deep photometric observations revealed the optical counterpart of the companion, thereby confirming its white-dwarf nature.

7.2 Questions and Thoughts for the Future

7.2.1 White Dwarf Physics

The white-dwarf companion to PSR J1909−3744 has a thick envelope and therefore suggests that white-dwarf models incorrectly produce hydrogen flashes for lower-than-0.2 $M_o dot$ stars. Indeed all white dwarfs studied herein have large ages—otherwise the possibility for observing them simultaneously is insignificant.

This obviously raises the question: What stops hydrogen flashes from occurring? Since the main driver for runaway fusion is gravitational settling of CNO nuclei, a likely possibility is that this process is somehow altered by rotational mixing. The solution on the other hand could be much more trivial, e.g. the standard "solar" metallicity of $z = 0.02$ adopted for these models could be overestimated.

I am currently trying to explore these effects in my models in a self-consistent manner. The initial results are promising but further investigation is required before drawing definite conclusions. To that end, high-resolution Echelle spectra may yield precise measurements of rotational velocities and provide an observational test of some of the proposed hypotheses.

7.2.2 *Millisecond Pulsar Ages*

On a related issue, accurate white-dwarf cooling curves will help infer precise ages and shed some (observational) light to the long-standing problem of Pulsar spin-down evolution and magnetic field decay. Hopefully, the ongoing deep Pulsar surveys carried out with Effelsberg, Arecibo, GBT and Parkes will yield a handfull of "J1909-like" systems to work with. Similarly, there is already a significant number of double-degenerate and detached binaries hosting low-mass white dwarfs (e.g. Kilic et al. 2012, and references therein) that can be used in a similar manner.

7.2.3 *Evolution of Low-Mass X-ray Binaries*

PSR J1738+0333 and PSR J0348+0432 are members of an emerging population of short-orbit binaries that no binary-evolution code can currently reproduce precisely. In particular, it is highly unclear how the donor star manages to detach from its Roche-lobe at such a short orbital period and why this is not the case for the progenitors of "black-widow" pulsars.

So far, some solutions to this problem (e.g. Podsiadlowski et al. 2002) have come by fine-tuning the free parameters involved in magnetic braking and mass loss contributions (see Eqs. 2.20 and 2.21). However, non of these are stable and hence not likely to reflect the underlying mechanism (since we know 5 of these systems already).

One possibility would (again) be that diffusion plays an important role at the last stages of RGB evolution. My colleagues and I hope to investigate this in more detail, along with other factors that might be important (e.g. irradiation or inhomogeneous mass loss).

7.2.4 *Neutron Star Masses*

The mass measurement of PSR J0348+0432 raises the question if heavy neutron stars are a significant fraction of the neutron star population. If this is the case then, in turn, one also wonders if the high mass can be a direct product of the supernova explosion or if it is a mere consequence recycling. Looking at the significantly different masses of PSR J1738+0333 and PSR J0348+0432 at presumably formed through the same evolutionary tempting to conclude that neutron stars can indeed be born massive. Further evidence for that comes from the original massive binary PSR J1614−2230, which probably evolved via a common envelope and thus had very little time to accrete (Tauris et al. 2011). On the other hand, there are many uncertainties that still need to be addressed. For example, an alternative explanation could be that the recycling mechanism (more specifically the accretion efficiency) is pulsar-specific. Indeed, the pulsars studied in this work have very different observational properties and it would be interesting to look for patterns in the statistics of future discoveries.

7.2.5 *Strong-Field Gravity*

The timing precision for PSRs J1738+0333 and J0348+0432 will continue to improve with time, resulting to more stringent constraints on their post-Keplerian parameters. Concerning Scalar–Tensor gravity, both these pulsar experiments will eventually surpass in sensitivity the Solar-system experiments for the entire range of the parameter space. To that end, some care must be taken in improving the masses, radii and velocity constraints presented here. These will become increasingly important for both subtracting the kinematic contributions and improving the strong-field-independent masses.

References

Bailes M., Bates S. D., Bhalerao V., Bhat N. D. R., Burgay M., Burke-Spolaor S., D'Amico N., Johnston S., Keith M. J., Kramer M., Kulkarni S. R., Levin L., Lyne A. G., Milia S., Possenti A., Spitler L., Stappers B., van Straten W., 2011, Science, 333, 1717.

Demorest P. B., Pennucci T., Ransom S. M., Roberts M. S. E., Hessels J. W. T., 2010, Nature, 467, 1081.

Kilic M., Thorstensen J. R., Kowalski P. M., Andrews J., 2012, MNRAS, 423, L132.

Podsiadlowski P., Rappaport S., Pfahl E. D., 2002, ApJ, 565, 1107.

Tauris T. M., Langer N., Kramer M., 2011, ArXiv e-prints.

Tremblay P.., Ludwig H.., Steffen M., Bergeron P., Freytag B., 2011, ArXiv:1302.2013

Tremblay P.-E., Ludwig H.-G., Steffen M., Freytag B., 2013.

Zeitfracht Medien GmbH
Ferdinand-Jühlke-Straße 7
99095 Erfurt, Deutschland
produktsicherheit@kolibri360.de